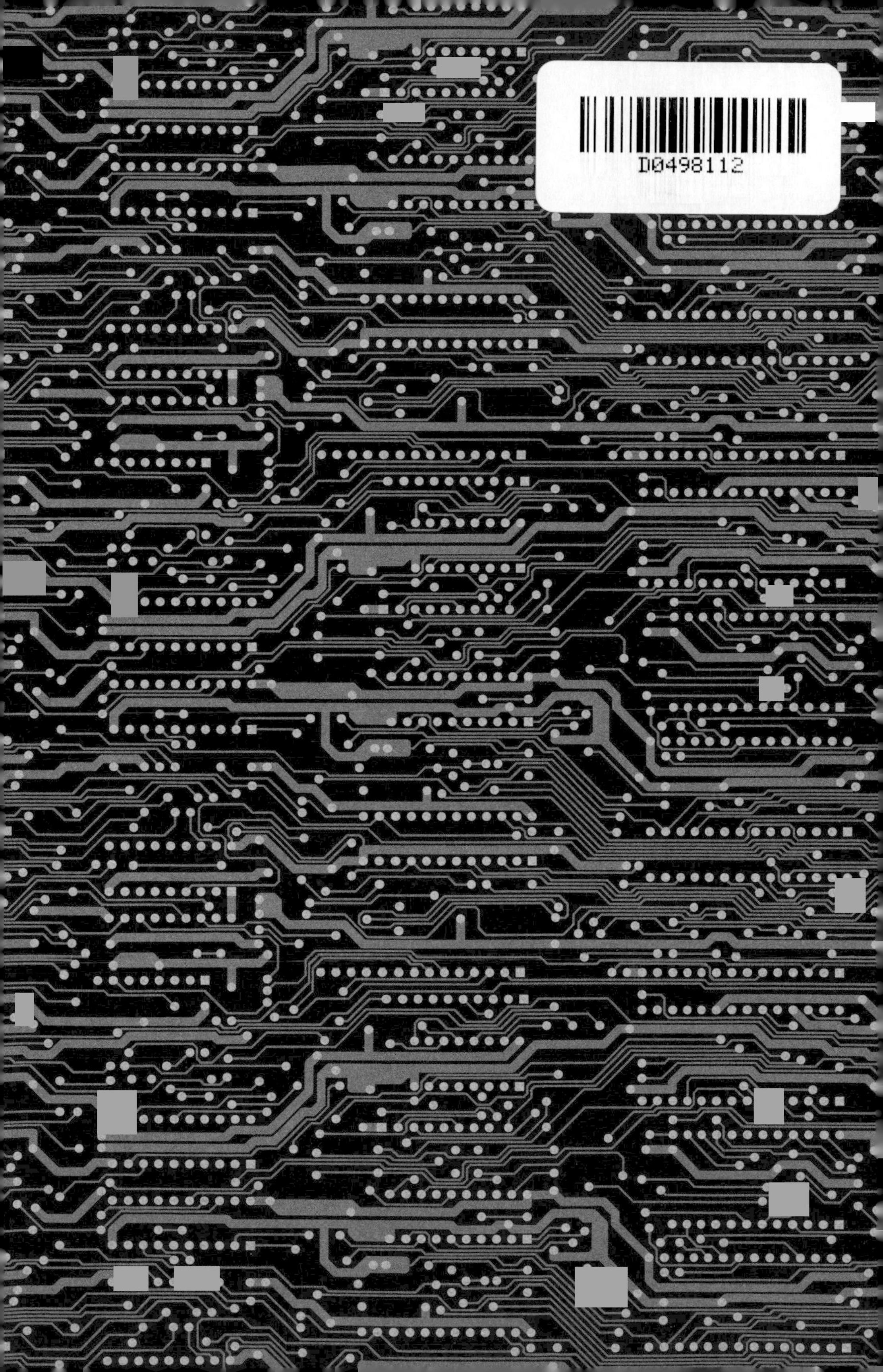

PRAISE FOR
The Hardware Hacker

"Hardware, says bunnie, is a world without secrets: if you go deep enough, even the most important key is expressed in silicon or fuses. bunnie's is a world without mysteries, only unexplored spaces. This is a look inside a mind without peer."
—Edward Snowden

"A tour de force that combines the many genius careers of one of the world's great hacker-communicators: practical, theoretical, philosophical, and often mind-blowing."
—Cory Doctorow, author of *Little Brother* and technology activist

"bunnie lives in the world of hardware where the solder meets the PCB. He has more practical experience and is a better teacher of how the ecosystem of hardware works than any other person I've ever met, and I know a lot of people in this space. He has rendered this experience and expertise into an amazing book—a hacker's-point-of-view bible to anyone trying to work in or understand and work in the emerging and evolving world of hardware."
—Joi Ito, Director, MIT Media Lab

"bunnie is the ultimate tour guide of hardware hacking as it stands today, with an eye toward the sublime art of how things are *really* made. *The Hardware Hacker* will take you on a journey through the factories of the world, covering both the technical and ethical implications of the 'stuff' we manufacture and buy."
—LIMOR "LADYADA" FRIED, FOUNDER & ENGINEER, ADAFRUIT INDUSTRIES

"Curious how the devices in our daily lives come into being? Want to manufacture your own project? In this well-written book, bunnie describes the ins and outs of the manufacturing process in China. A very entertaining and informative read."
—MITCH ALTMAN, INVENTOR OF TV-B-GONE®

"*The Hardware Hacker* is, at its core, the primer for understanding the culture of making something in China, how to build thousands of things, and why Open Hardware works."
—HACKADAY

The Hardware Hacker

Adventures in Making and Breaking Hardware

Andrew “bunnie” Huang

Printed in USA

First printing
20 19 18 17 1 2 3 4 5 6 7 8 9

ISBN-10: 1-59327-758-X
ISBN-13: 978-1-59327-758-1

Publisher: William Pollock
Production Editor: Alison Law
Cover and Jacket Design: Hotiron Creative
Interior Design: Beth Middleworth
Developmental Editor: Jennifer Griffith-Delgado
Copyeditor: Rachel Monaghan
Compositor: Alison Law
Proofreader: Emelie Burnette
Indexer: BIM Creatives, LLC.

For information on distribution, translations, or bulk sales, please contact No Starch Press, Inc. directly:

No Starch Press, Inc.
245 8th Street, San Francisco, CA 94103
phone: 1.415.863.9900; info@nostarch.com; www.nostarch.com

Library of Congress Cataloging-in-Publication Data

Names: Huang, Andrew, author.
Title: The hardware hacker : adventures in making and breaking hardware / Andrew "Bunnie" Huang.
Description: 1st ed. | San Francisco : No Starch Press, Inc., [2017]
Identifiers: LCCN 2016038846 (print) | LCCN 2016049285 (ebook) | ISBN 9781593277581 (pbk.) | ISBN 159327758X (pbk.) | ISBN 9781593278137 (epub) | ISBN 1593278136 (epub) | ISBN 9781593278144 (mobi) | ISBN 1593278144 (mobi)
Subjects: LCSH: Electronic apparatus and appliances--Design and construction. | Electronic apparatus and appliances--Technological innovations. | Computer input-output equipment--Design and construction. | Reverse engineering. | Electronic industries. | Huang, Andrew.
Classification: LCC TK7836 .H83 2017 (print) | LCC TK7836 (ebook) | DDC 621.381092--dc23
LC record available at https://lccn.loc.gov/2016038846

To all the wonderful, patient, and accepting people
who have supported this eccentric hacker

ACKNOWLEDGMENTS

Thanks to all the hard-working staff at No Starch Press for making this book happen. In particular, thanks to Bill Pollock for conceiving and sponsoring the effort, and thanks to Jennifer Griffith-Delgado for compiling, editing, and arranging my writing into the form of this book.

brief contents

preface xvii

part 1
adventures in manufacturing 1

chapter 1. made in china 7

chapter 2. inside three very different factories 43

chapter 3. the factory floor 73

part 2
thinking differently:
intellectual property in china 115

chapter 4. gongkai innovation 119

chapter 5. fake goods 143

part 3
what open hardware means to me 175

chapter 6. the story of chumby 181

chapter 7. novena: building my own laptop 215

chapter 8. chibitronics: creating circuit stickers 251

part 4
a hacker's perspective 275

chapter 9. hardware hacking 279

chapter 10. biology and bioinformatics 325

chapter 11. selected interviews 357

epilogue 383

index 384

contents in detail

preface xvii

part 1 adventures in manufacturing 1

1. made in china 7

The Ultimate Electronic Component Flea Market 8
The Next Technological Revolution 14
Touring Factories with Chumby 16
- Scale in Shenzhen 17
- Feeding the Factory 18
- Dedication to Quality 20
- Building Technology Without Using It 23
- Skilled Workers 24
- The Need for Craftspeople 26
- Automation for Electronics Assembly 29
- Precision, Injection Molding, and Patience 31
- The Challenge of Quality 34

Closing Thoughts 42

2. inside three very different factories 43

Where Arduinos Are Born 44
- Starting with a Sheet of Copper 46
- Applying the PCB Pattern to the Copper 49
- Etching the PCBs 51
- Applying Soldermask and Silkscreen 53
- Testing and Finishing the Boards 54

Where USB Memory Sticks Are Born 57
The Beginning of a USB Stick 57
Hand-Placing Chips on a PCB 59
Bonding the Chips to the PCB 61
A Close Look at the USB Stick Boards 61
A Tale of Two Zippers 64
A Fully Automated Process 67
A Semiautomated Process 68
The Irony of Scarcity and Demand 70

3. the factory floor 73

How to Make a Bill of Materials 74
A Simple BOM for a Bicycle Safety Light 74
Approved Manufacturers 76
Tolerance, Composition, and Voltage Specification 76
Electronic Component Form Factor 77
Extended Part Numbers 78
The Bicycle Safety Light BOM Revisited 79
Planning for and Coping with Change 82
Process Optimization: Design for Manufacturing 84
Why DFM? 85
Tolerances to Consider 86
Following DFM Helps Your Bottom Line 88
The Product Behind Your Product 91
Testing vs. Validation 97
Finding Balance in Industrial Design 100
The chumby One's Trim and Finish 101
The Arduino Uno's Silkscreen Art 104
My Design Process 105
Picking (and Maintaining) a Partner 107
Tips for Forming a Relationship with a Factory 107
Tips on Quotations 108
Miscellaneous Advice 111
Closing Thoughts 113

part 2
thinking differently:
intellectual property in china **115**

4. gongkai innovation **119**

I Broke My Phone's Screen, and It Was Awesome 120
Shanzhai as Entrepreneurs 121
Who Are the Shanzhai? 122
More Than Copycats 123
Community-Enforced IP Rules 124
The $12 Phone 126
Inside the $12 Phone 128
Introducing Gongkai 131
From Gongkai to Open Source 134
Engineers Have Rights, Too 135
Closing Thoughts 141

5. fake goods **143**

Well-Executed Counterfeit Chips 143
Counterfeit Chips in US Military Hardware 149
Types of Counterfeit Parts 150
Fakes and US Military Designs 153
Anticounterfeit Measures 154
Fake MicroSD Cards 156
Visible Differences 157
Investigating the Cards 158
Were the MicroSD Cards Authentic? 159
Further Forensic Investigation 160
Gathering Data 162
Summarizing My Findings 166
Fake FPGAs 168
The White Screen Issue 168
Incorrect ID Codes 170
The Solution 172
Closing Thoughts 174

part 3 what open hardware means to me 175

6. the story of chumby 181

A Hacker-Friendly Platform 182
Evolving chumby 184
  A More Hackable Device 186
  Hardware with No Secrets 187
The End of Chumby, New Adventures 189
Why the Best Days of Open Hardware Are Yet to Come 205
  Where We Came From: Open to Closed 206
  Where We Are: "Sit and Wait" vs. "Innovate" 208
  Where We're Going: Heirloom Laptops 210
  An Opportunity for Open Hardware 211
Closing Thoughts 214

7. novena: building my own laptop 215

Not a Laptop for the Faint of Heart 217
Designing the Early Novena 219
  Under the Hood 219
  The Enclosure 224
The Heirloom Laptop's Custom Wood Composite 227
  Growing Novenas 228
  The Mechanical Engineering Details 229
Changes to the Finished Product 232
  Case Construction and Injection-Molding Problems 233
  Changes to the Front Bezel 237
  DIY Speakers 238
  The PVT2 Mainboard 238
  A Breakout Board for Beginners 241
  The Desktop Novena's Power Pass-Through Board 242
  Custom Battery Pack Problems 243
  Choosing a Hard Drive 244
  Finalizing Firmware 246
Building a Community 247
Closing Thoughts 249

8. chibitronics: creating circuit stickers 251

Crafting with Circuits 257
- Developing a New Process 259
- Visiting the Factory 260
- Performing a Process Capability Test 261

Delivering on a Promise 264
Why On-Time Delivery Is Important 266
Lessons Learned 266
- Not All Simple Requests Are Simple for Everyone 267
- Never Skip a Check Plot 268
- If a Component Can Be Placed Incorrectly, It Will Be 268
- Some Concepts Don't Translate into Chinese Well 270
- Eliminate Single Points of Failure 271
- Some Last-Minute Changes Are Worth It 271
- Chinese New Year Impacts the Supply Chain 272
- Shipping Is Expensive and Difficult 273
- You're Not Out of the Woods Until You Ship 274

Closing Thoughts 274

part 4
a hacker's perspective 275

9. hardware hacking 279

Hacking the PIC18F1320 281
- Decapping the IC 282
- Taking a Closer Look 283
- Erasing the Flash Memory 284
- Erasing the Security Bits 285
- Protecting the Other Data 287

Hacking SD Cards 289
- How SD Cards Work 290
- Reverse Engineering the Card's Microcontroller 293
- Potential Security Issues 298
- A Resource for Hobbyists 298

Hacking HDCP-Secured Links to Allow Custom Overlays 298
- Background and Context 300
- How NeTV Worked 302

Hacking a Shanzhai Phone 306
The System Architecture 306
Reverse Engineering the Boot Structure 311
Building a Beachhead 315
Attaching a Debugger 317
Booting an OS 321
Building a New Toolchain 321
Fernvale Results 323
Closing Thoughts 324

10. biology and bioinformatics 325

Comparing H1N1 to a Computer Virus 327
DNA and RNA as Bits 328
Organisms Have Unique Access Ports 330
Hacking Swine Flu 331
Adaptable Influenza 333
A Silver Lining 335
Reverse Engineering Superbugs 335
The O104:H4 DNA Sequence 336
Reversing Tools for Biology 338
Answering Biological Questions with UNIX Shell Scripts 340
More Questions Than Answers 342
Mythbusting Personalized Genomics 344
Myth: Having Your Genome Read Is Like Hex-Dumping the ROM of Your Computer 344
Myth: We Know Which Mutations Predict Disease 345
Myth: The Reference Genome Is an Accurate Reference 345
Patching a Genome 346
CRISPRs in Bacteria 347
Determining Where to Cut a Gene 350
Implications for Engineering Humans 351
Hacking Evolution with Gene Drive 352
Closing Thoughts 354

11. selected interviews **357**

Andrew "bunnie" Huang: Hardware Hacker (CSDN) 357

About Open Hardware and the Maker Movement 358

About Hardware Hackers 367

The Blueprint Talks to Andrew Huang 372

epilogue **383**

index **384**

preface

When Bill Pollock, founder of No Starch Press, first contacted me with the idea of publishing a compilation of my writings, I was skeptical. I didn't think there would be enough material to fill a hundred pages. It seems I was wrong.

My mother often said, "It doesn't matter what's in your head if you can't tell people what's in it," and when I was in seventh grade, she enrolled me in an after-school essay writing class. I hated the class at the time, but in retrospect, I'm thankful. Starting with my college application essays and up to this day, I've found the ability to organize my thoughts into prose invaluable.

Most of the material in this book was originally published on my blog, but as you'll soon see, those posts weren't puff pieces written to drive ad revenue. One reason I write is to solidify my own understanding of complicated subjects. It's easy to believe you understand a topic until you try to explain it to someone else in a rigorous fashion. Writing is how I distill my intuition into structured knowledge; I only write when I find something interesting to write about, and then I post it with a CC BY-SA license to encourage others to share it.

This book includes a selection of my writings on manufacturing, intellectual property (with a focus on comparing Western versus Chinese perspectives), open hardware, reverse engineering, and biology and bioinformatics. The good editors at No Starch Press also curated a couple of interviews I've done in the past that were particularly informational or insightful. The common thread throughout these diverse topics is hardware: how it's made, the legal frameworks around it, and how it's unmade. And yes, biological systems are hardware.

I've always gravitated toward hardware because while I'm not particularly gifted when it comes to abstract thought (hence the need to write to organize my thoughts), I am pretty good with my hands. I have a much better chance of understanding things that I can see with my own two eyes.

My entire understanding of the world has always been built on a series of simple, physical experiences, starting from when I stacked blocks and knocked them over as a child. This book shares some of my more recent experiences. I hope that by reading them, you will gain a deeper understanding of the world of hardware, without having to spend decades stacking blocks and knocking them over.

Happy hacking,
—b.

Part 1

adventures in manufacturing

I first set foot in China in November 2006. I had no idea what I was walking into. When I told my mother I was going to visit Shenzhen, she exclaimed, "Why are you going there? It's just a fishing village!" She wasn't wrong: Shenzhen was just a town of 300,000 back in 1980, but it had exploded into a megacity of 10 million in less than 30 years. Between my first visit and the time I wrote this book, Shenzhen gained an estimated 4 million people—more than the population of Los Angeles.

In a way, my understanding of manufacturing over the years has mirrored Shenzhen's growth. Before going to China, I had never mass-produced anything. I didn't know anything about supply chains. I had no idea what "operations and logistics" meant. To me, it sounded like something out of a math or programming textbook.

Still, Steve Tomlin, my boss at the time, charged me with figuring out how to build a supply chain suitable for our hardware startup, Chumby. Sending a novice into China was a big risk, but my lack of preconceived notions was more of an asset than a liability. Back then, venture capitalists shunned hardware, and China was only for established companies looking to build hundreds of thousands of units of a given product. My first set of tours in China certainly supported that notion, as I primarily toured mega-factories serving the *Fortune* 500.

Chumby was lucky to be taken under the wing of PCH International as its first startup customer. At PCH, I was mentored by some of the finest engineers and supply chain specialists. I was also fortunate to be allowed to share my experiences on my blog, as Chumby was one of the world's first open hardware startups.

Although meeting the minimum order volumes of our conventional manufacturing partners was a constant struggle, I kept noticing small things that didn't square with conventional wisdom. Somehow, local Chinese companies were able to remix technology into boutique products. The so-called shanzhai integrated cell phones into all kinds of whimsical forms, from cigarette lighters to ornamental golden Buddha statuettes (more on this in Chapter 4). The niche nature of these products meant they had to be economical to produce in smaller volumes. I also noticed that somehow factories were able to rapidly produce bespoke adapter circuits and testing apparatuses of surprisingly high quality in single-unit volumes. I felt there was more to the ecosystem—a story that was being told over and over again—but few had the time to listen, and those who did heard only the parts they wanted to hear.

The financial crisis of 2008 changed everything. The consumer electronics market was crushed, and factories that were once too busy printing money were now swimming in excess capacity. I made friends at several medium-sized factories in the area. I started to inquire about how, exactly, these factories were able to so nimbly produce their internal test equipment, and how shanzhai were able to prototype and build such bespoke phones.

The bosses and engineers were initially reticent, not because they wanted to hide potential competitive advantages from me, but because they were ashamed of their practices. Foreign clients were full of corporate process, documentation, and quality procedures, but they also paid dearly for such overhead. Local companies were much more informal and pragmatic. So what if a bin is labeled "scrap"? If the bits inside are suitable for a job, then use them!

I wanted in. As an engineer, tinkerer, and hacker, I cared a lot about the cost to produce a few units, and a couple of minor assembly defects was nothing compared to the design issues I had to debug. I eventually managed to coax a factory into letting me build a part using its low-quality but ultra-cheap assembly process.

The trick was to guarantee that I would pay for all the product, including defective units. Most customers refuse to pay for imperfect goods, forcing the factory to eat the cost of any part that isn't exactly to specification. Thus, factories strongly dissuade customers from using cheaper but low-quality processes.

Of course, my promise to pay for defective product meant there was no incentive for the factory to do a good job. It could have, in theory, just handed me a box of scrap parts and I'd still have had to pay for it. But in reality, nobody had such ill intentions; as long as everyone simply tried their best, they got it right about 80 percent of the time. Since small-volume production costs are dominated by setup and assembly, my bottom line was still better despite throwing away 20 percent of my parts, and I got parts in just a couple of days instead of a couple of weeks.

Having options to trade cost, schedule, and quality against each other changes everything. I've made it a point to discover more alternative production methods and continue shortening the path between ideas and products, with ever more options along the cost-schedule-quality spectrum.

After Chumby, I decided to remain unemployed, partly to give myself time for discovery. For example, every January, instead of going to the frenzied Consumer Electronics Show (CES) in Las Vegas, I rented a cheap apartment in Shenzhen and engaged in the "monastic study of manufacturing"; for the price of one night in Las Vegas, I lived in Shenzhen for a month. I deliberately picked neighborhoods with no English speakers and forced myself to learn the language and customs to survive. (Although I'm ethnically Chinese, my parents prioritized accent-free fluency in English over learning Chinese.) I wandered the streets at night and observed the back alleys, trying to make sense of all the strange and wonderful things I saw going on during the daytime. Business continues in Shenzhen until the wee hours of the morning, but at a much slower pace. At night, I could make out lone agents acting out their interests and intentions.

If there's one thing those studies taught me, it's that I have a lot more to learn. The Pearl River Delta ecosystem is incomprehensibly vast. As with the Grand Canyon, simply hiking one trail from rim to base doesn't mean you've seen it all. I have, however, picked up enough knowledge to build a custom laptop and to develop a new process for peel-and-stick electronic circuits.

In this part of the book, you'll follow my journey as I learned the Shenzhen ecosystem over the years, via a remix of blog posts that I wrote along the way. Some of the essays are reflections on particular aspects of Chinese culture; others are case studies of specific manufacturing practices. I conclude with a chapter called "The Factory Floor," a set of summary recommendations for anyone considering outsourced manufacturing. If you're in a hurry, you can skip all the background and go directly there.

However, hindsight is 20/20. Once you've walked a path, it's easy to point out the shortcuts and hazards along the way; it's even easier to forget all of the wrong turns and bad assumptions. There's no one-size-fits-all method for approaching China, and my hope is that by reading these stories, you can come to your own (perhaps different) conclusions that better serve your unique needs.

1. made in china

Before my first visit to China, I was convinced that Akihabara in Tokyo was the go-to place for the latest electronics, knick-knacks, and components. That changed in January 2007, when I first set eyes on the SEG Electronics Market in Shenzhen. SEG is eight floors of all the components a hardware addict could ever want, and only later did I learn that it's just the tip of the Hua Qiang electronics district iceberg.

As the lead hardware engineer at Chumby at the time, I was in China with then-CEO Steve Tomlin to figure out how to make chumbys (an open source, Wi-Fi-enabled content delivery device) cheaply and on time. With prices like those at SEG, we were definitely in the right country to make at least the first part of that mission a success.

Shenzhen's SEG Electronics Market, the new electronics mecca. Akihabara, eat your heart out!

THE ULTIMATE ELECTRONIC COMPONENT FLEA MARKET

When I first stepped into the SEG building, I was assaulted by a whirlwind of electronic components: tapes and reels of resistors and capacitors, ICs of every type, inductors, relays, pogo pin test points, voltmeters, and trays of memory chips. As a total newcomer to manufacturing in volume, I was blown away by everything I saw at SEG.

All of those parts were crammed into tiny six-by-three-foot booths, each with a storekeeper poking away at a laptop. Some storekeepers played *Go*, and some counted parts. Some booths were true mom-and-pop shops, with mothers tending to babies and kids playing in the aisles.

A couple of family-run component shops

Other booths were professional setups with uniformed staff, and these worked like a bar—complete with stools—for electronic components.

A swanky professional parts seller

No one at SEG says, "Oh, you can get 10 of these LEDs or a couple of these relays," like you might hear in Akihabara. No, no. These booths specialize, and if you see a component you like, you can usually buy several tubes, trays, or reels of it; you can get enough to go into production the next day.

Looking around the market, I saw a woman sorting stacks of 1GB mini-SD cards like poker chips. A man was putting sticks of 1GB Kingston memory into retail packages, and next to him, a girl was counting resistors.

The bottom-left corner of this display was packed with all kinds of SD cards.

Another booth had stacks of power supplies, varistors, batteries, and ROM programmers, and yet another had chips of every variety: Atmel, Intel, Broadcom, Samsung, Yamaha, Sony, AMD, Fujitsu, and more. Some chips were clearly ripped out of used equipment and remarked, some of them in brand-new laser-marked OEM packaging.

The sheer quantity of chips for sale at a single booth at SEG was incredible.

I saw chips that I could never buy in the United States, reels of rare ceramic capacitors that I could only dream about at night. My senses tingled; my head spun. I couldn't suppress a smile of anticipation as I walked around the next corner to see shops stacked floor to ceiling with probably 100 million resistors and capacitors.

Reels and reels of components, in every shop window

Sony CCD and CMOS camera elements! I couldn't buy those in the United States if I pulled teeth out of the sales reps. (Some sellers even have the datasheets behind the counter; always ask.) Next, I spotted a stack of Micrel regulator chips, followed by a Blackfin DSP chip for sale. Nearby, a lady counted 256Mb DRAM chips—trays of 108 components, stacked 20 high, in perhaps 10 rows.

The equivalent of Digi-Key's entire stock of DRAM chips sat right in front of me!

And across from her were a half-dozen more little shops packed with chips just like hers. At one shop, a man stood proudly over a tray of 4Gb NAND flash chips. All of this was available for a little haggling, a bit of cash, and a hasty good-bye.

A close look at a tray of 4Gb flash chips

And that's just the first two floors of SEG. There are six more floors of computer components, systems, laptops, motherboards, digital cameras, security cameras, thumb drives, mice, video cameras, high-end graphics cards, flat-panel displays, shredders, lamps, projectors—you name it. On weekends, "booth babes" dressed in outrageous Acer-branded glittery bodysuits loiter around, trying to pull you in to buy their wares. This market has all the energy of a year-round CES meets Computex, except instead of just showing off the latest technology, the point is getting you into these booths to buy that hardware. Trade shows always feel like a bit of a strip tease, with your breath making ghostly rings on the glass as you hover over the unobtainable wares underneath.

But SEG is no strip tease. It's the orgy of consumer and industrial electronic purchasing, where you can get your grubby paws on every piece of equipment for enough *kuai** out of your wallet. Between the smell, the bustle, and the hustle, SEG is the ultimate electronic component flea market. It's as if Digi-Key went mad and let monkeys into its Minnesota warehouse, and the resulting chaos spilled into a flea market in China.

Of course, a lot of the parts I marveled at in 2007 are antiques now. For example, 4Gb flash chips are trash, and 1GB flash disks are old news. At the time, however, those things were a big deal, and SEG is still the best place to get the latest tech in bulk.

THE NEXT TECHNOLOGICAL REVOLUTION

Three blocks down the street from SEG lay the Shenzhen Bookstore.† The first and most visible rack was a foreign book section, packed with classics like Stanford University professor Thomas Lee's *The Design of CMOS Radio-Frequency Integrated*

* Colloquial word for *yuan*, the base counting unit for the *renminbi (RMB)*, the currency in China.

† This bookstore has closed since the visit I describe here.

Circuits and several titles by UCLA professor Behzad Razavi. I picked up Lee's book, and it cost 68 kuai, or $8.50 USD. Holy cow! Jin Au Kong's book on Maxwell's equations? $5. Jin Au Kong *taught* me Maxwell's equations at MIT.

I went on a spree, packing my bag with six or seven titles, probably around $700 worth of books if I'd bought them in the United States. At the checkout counter, I bought them for less than $35, complete with the supplemental CDs, saving about $665. That's equivalent to buying an economy-class ticket to Hong Kong!

In China, knowledge is cheap. Components are cheap. The knowledge in the books at the Shenzhen Bookstore was the Real Deal, the parts to use that knowledge are down the street at SEG, and within an hour's drive north are probably 200 factories that can take any electronics idea and pump it out by the literal boatload. These are no backward factories, either. With my own eyes, I saw name-brand, 1,550-nanometer, single-mode, long-haul, fiber-optic transceivers being built and tested there. Shenzhen is fertile ground, and you need to see it to understand it.

Shenzhen has the pregnant feel of the swapfests in Silicon Valley back in the '80s, when all the big companies were just being founded and starting up, except magnified by 25 years of progress in Moore's law and the speed of information flow via the internet. In this city of 12 million people, most are involved in tech or manufacturing, many are learning English, and all of them are willing to work hard.

There has to be a Jobs and Wozniak there somewhere, quietly building the next revolution. But I'm a part of Shenzhen, too, and I still tremble in my boots with terror and excitement at the thought of being part of that revolution. This is my story, starting with that eye-opening trip to Shenzhen for Chumby.

TOURING FACTORIES WITH CHUMBY

In September 2006, Chumby was just a team of about a half-dozen people, and we had just given away about 200 early prototype chumby devices at FOO Camp, a conference put on by Tim O'Reilly. The devices were well received by the FOO Camp attendees, so I got the go-ahead to build the Asian supply chain.

Steve and I went to China to visit potential factories in November, but before we left, we had a trusted vendor in the United States give their best price for the job as a baseline for negotiations with the Chinese manufacturers. Then, we called up a lot of friends with experience in China and lined up about six factory tours. We hit quite a variety of places, from specialty factories as small as 500 people to mega-factories with over 40,000 people.

There's no substitute for going to China to tour a factory. Pictures can only tell the story framed by the photographer, and you can't get a sense of a facility's scale and quality without seeing it firsthand. In general, factories welcome you to take a tour, and I wouldn't work with one that didn't allow me to visit. However, most factories do appreciate a week's notice, although as your relationship with them progresses, things should become more open and transparent.

Speaking of openness, Chumby's open source nature helped the factory selection process a lot. First, we had no fears about people stealing our design (we were giving it away already), so we'd eliminated the friction of NDAs (non-disclosure agreements) when sharing critical information like the bill of materials. I think this gave us a better reception with factories in China; they seemed more willing to open up to us because we were willing to open up to them. Second, there was no question in any factory's mind that this was a competitive situation. Anybody could and would quote and bid on our job (in fact, we received a few unsolicited quotations that were quite competitive), so it saved a round of huffing and puffing.

After reviewing several manufacturing options, Steve and I eventually decided to work with a company called PCH China Solutions. PCH itself owns only a few facilities, but it has a comprehensive network of trusted and validated vendors, primarily in China but also in Europe and the United States. Not surprisingly, the factories that PCH subcontracts to were some of the best facilities we visited in China. PCH is actually headquartered out of Ireland—thus most of their staff engineers are Irish—so there was also no language barrier for us. (PCH engineers are also hardworking, resourceful, and well trained—and, as a bonus, they always seem to know the best place to find a pint, no matter where they are. I had no idea China had so many Guinness taps!)

There's a lot to take in when you tour even one factory, let alone a half-dozen, and it's easy to get overwhelmed and lost in the vagaries of electronics manufacturing. But there were some key details I found most fascinating during my factory tours for Chumby and in working with PCH to bring the chumby to life.

Scale in Shenzhen

One stunning thing about working in China is the sheer scale of the place. I haven't been to an auto plant in Michigan or to the Boeing plant in Seattle, but I get the sense that Shenzhen gives both a run for their money in terms of scale. In 2007, Shenzhen had 9 million people.

To give you an idea of the scale of a Shenzhen factory, the New Balance factory there employed 40,000 people and had the capacity to produce over a million shoes a month. I estimate that from raw fabric to finished shoe, the process took about 50 minutes, and every perfectly stitched bundle of plastic and leather was sewn by hand on an industrial sewing machine. The stations are designed so that each stage in the process takes a worker about 30 seconds.

Of course, the New Balance factory is dwarfed by Foxconn, the factory where iPods and iPhones are made.

You know you're big when you have your own exit off the freeway.

Foxconn is a huge facility, apparently with over 250,000 employees, and it has its own special free trade status. The entire facility is walled off, and I've heard you need to show your passport and clear customs to get into the facility. That's just short of the nuclear-powered robotic dogs from the nation-corporation franchulates of Neal Stephenson's *Snow Crash*.

Feeding the Factory

There's an old Chinese saying: *min yi shi wei tian*. A literal translation would be "people consider food divine" or "for people, food is next to heaven." You can also look at it as a piece of governing advice: "the government's mandate [synonymous with heaven] is only as robust as the food on people's plates." Or, you can interpret it as an excuse to procrastinate: "let's eat first [since it is as important as heaven]."

Whichever way you cut it, I think the saying still holds in China. One important metric for gauging how well a factory treats its employees is how good the food is, as it's common for factory workers to be housed, fed, and cared for on site.

The food is actually quite good at some factories. For example, when eating with the workers at the factory that manufactured chumby circuit boards, I was served a mix of steamed fish, broiled pork, egg rolls, clean fried vegetables, and some pickled-vegetable-and-meat combo. Rice, soup, and apples were also provided in "help yourself" quantities.

A meal from the factory that made the chumby circuit boards

Every facility I visited also had separate utensils and plates for guests. At one factory, my food was served on a Styrofoam plate with disposable chopsticks, while a factory worker I ate with was served food on a steel plate with steel chopsticks. I hadn't passed the factory's physical examination, so they gave me disposable eating tools to prevent me from contaminating the factory with potential foreign diseases.

Going back to scale, some factory food operations are impressively large. I heard that Foxconn's workers consume 3,000 pigs a day. From pigs to iPhones, it all happens right here in Shenzhen!

A truckload of pigs, exiting the highway toward Foxconn

Dedication to Quality

After I started working with PCH on actually manufacturing the chumby, I ran into a situation sometime around June 2007 that showed me just how dedicated the factory workers in Shenzhen were to getting their jobs right.

I had updated the chumby motherboard to include an electret microphone, with an integral pre-amp field-effect transistor (FET). The microphone needed to be inserted in the correct orientation with respect to the circuit so the FET would receive a proper bias current.

The first samples I got back from PCH's factory had the microphone in backward, and I called the factory to tell them to reverse its polarity. I was going to visit the factory the next week, and I wanted to see corrected samples. When I arrived and tested the microphone, I found to my dismay that the microphones were *still* not working.

How could that be? There are only two ways to connect a microphone.

It turns out there were two operators on the line assembling the microphone. One soldered the red and black wires to the microphone. The next soldered these red and black wires to the circuit board. The operators were told to reverse the order, and both of them dutifully complied—giving me a microphone that was still soldered in backward, but with the color of the wires swapped. (This is actually a pretty typical story for problems in China.)

The factory was scheduled to manufacture a first pilot run of 450 circuit boards the next day. Everything had to go perfectly for Chumby's production timeline to stay on schedule. We had soldering stencils rebuilt (we were debugging a yield issue with the QFN packaged audio CODEC as well) and ready by around noon, and by around 6 PM, I had the first boards in my hands to test. I ran the final factory test, and the device failed again—at the microphone. This was not a happy moment for anybody in the factory, as the factory was liable for any manufacturing defects.

I donned my smock and marched onto the line to start debugging the problem.

For the rest of the night, I remained in the factory, and so did every manager and tech involved in manufacturing the chumby. The pressure was enormous: right next to us was a line churning out 450 potentially defective circuit boards, and I was unwilling to pull the plug because I still didn't know what the root cause was, and we had to stay on schedule.

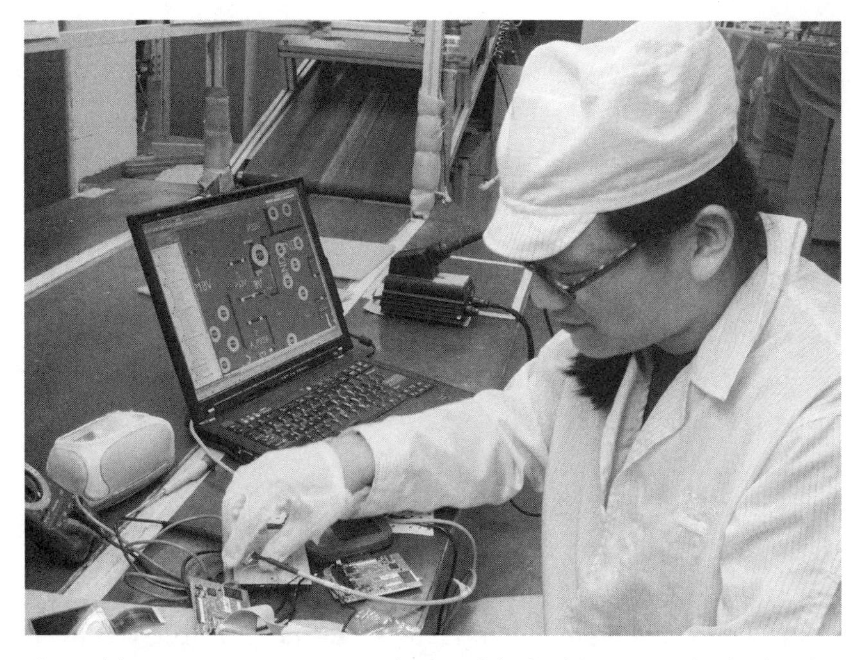

I was debugging circuits at 3 AM on the day of the final factory test for the chumby.

I literally had a panel of factory workers standing by the entire night to bring me anything I needed: soldering irons, test equipment, more boards, X-ray machines, microscopes. Remarkably, not a single person hesitated; not a single person complained; not a single person lost focus on the problem. People canceled dinner plans with friends without batting an eyelash. Anyone who wasn't needed in a particular moment was busy overseeing other aspects of the project. I hadn't seen blind dedication like that since I worked with the autonomous underwater robotics team at MIT.

And this went on until 3 AM.

Embarrassingly, the problem wasn't PCH's fault in the end. The problem was the new firmware release I received earlier that day from the team in the United States. It had a bug that disabled the microphone due to a hack that was accidentally checked into the build tree.

Even more impressively, when PCH found out, nobody was angry, and nobody complained. (Well, the saleswoman gave me a hard time, but I deserved it; she had been kind enough to accompany me on the production line all night long and be my translator, since my Mandarin wasn't up to snuff.) They were simply relieved that it wasn't their fault.

We all parted ways, and I came back into the factory the next day at 11 AM, after a good night's sleep. I saw Christy, the factory's project manager for manufacturing the chumby boards. I asked her when she came into work, and she told me she always has to report by 8 AM. I started to feel really bad; Christy stayed up late because of our bug, and she came in early while I slept in. I asked her why she stayed up so late even though she knew she had to report to work at 8 AM. She could have gone home, and we could have continued the next day.

She just smiled and said, "It's my job to make sure this gets done, and I want to do a good job."

Building Technology Without Using It

Here's another interesting story. On our way out of the factory floor one day, Xiao Li (the quality assurance manager at the factory where we made the chumby) asked me, "What does a chumby do?" I didn't speak Chinese very well, and she didn't speak English very well either, so I decided to start with a few basic questions.

I asked her if she knew what the World Wide Web was. She said no.

I asked her if she knew what the internet was. She said no.

I was stunned, and I didn't know what to say. How do you describe the color blue to the blind?

Xiao Li was an expert in building and testing computers. On some projects, she probably built PCs and booted Windows XP a hundred thousand times over and over again. (God knows

I heard that darn startup sound a zillion times during the microphone incident, as there was a bank of final test stations for ASUS motherboards right next to me.) But she didn't know what the internet was.

I had assumed that if you touched a computer, you were also blessed by the bounties of the internet. All at once, I felt like a spoiled snob and a pig for forgetting that Xiao Li probably couldn't afford a computer, much less broadband internet access. Given the opportunity, she was certainly smart enough to learn it all, but she was too busy making money that she probably sent back home to her family.

In the end, the best I could do was to tell Xiao Li that the chumby was a device for playing games.

Skilled Workers

Shenzhen workers may not know a lot about everything they make, but on top of their dedication, they are highly skilled. I once watched a guy working at the same factory that sewed the chumby bags, and I swear, he could sew cosmetic cases together at a rate of 5 seconds per bag. And he wasn't even 100 percent focused on his task; he was listening to his iPod while he sewed.

And apparently, he wasn't their fastest employee! They had someone about twice as fast, and he'd been with the company for about seven years. I went to watch the faster worker, but he had already gone to lunch because he'd finished everything; there were two enormous bins of finished cosmetic cases next to his workstation.

On a similar note, I was amazed to learn how rubberized tags (the ones you see all over clothes) are made in China. I always thought they were pressed by a machine, but I was wrong. All those words, colors, and letters are drawn by hand. Someone just places a logo stencil over the blank tag, paints over the stencil with amazing precision, and moves on to the next tag in their queue. When there are multiple colors, there's a person for each color, to keep the process quick.

I asked PCH if they had any mechanized factories for stuff like that. They told me the facilities exist, but the minimum order quantity is enormous (hundreds of thousands, sometimes millions) because of the extraordinarily low cost of the product and the relatively high cost of tooling for the automated process. This is consistent with what I've heard about McDonald's Happy Meal toys. They're usually held together with screws because it's cheaper to pay someone to screw together a toy over the whole production run than it is to make a steel injection-molding tool with the tolerances necessary for snapping the toys together.*

There was a similar trade-off inside the chumby hardware. There were four connectors on the internal chumby electronics. Using the US-based vendors that I could source, one connector had a best price of about $1 USD, and the other three had a best price of about $0.40 each. PCH's very talented sourcing expert (her reputation was feared and respected by every vendor) managed to find me connectors that cost $0.10 and $0.06, respectively, saving almost a full $2 in cost. There's one catch: the connectors lacked the sacrificial plastic pick-and-place pad that would enable them to be machine-assembled.

The solution? A person, of course.

* Due to high wage inflation since this particular visit, this is probably no longer true.

This man hand-placed the cheaper connectors on every chumby, for about a nickel per unit. Thanks to him, chumbys were $2 cheaper, which freed up more money for us consumers to spend at Starbucks.

The Need for Craftspeople

I'd like to introduce you to a man I know simply as Master Chao. I met him during the chumby manufacturing process, and I'm pretty sure that in your lifetime, you have used or seen something that he created.

When I went to the sample room for the factory where Master Chao worked, I was shocked at how many items on their shelf I had purchased, used, or seen in a store in the United States myself. Top-tier consumer brands manufacture their stuff in this factory, and to the best of my knowledge, the factory had just one master pattern maker at the time: Master Chao. He's had a hand in creating cosmetic bags for Braun, accessory cases for Microsoft, and the medical braces for major brands sold in drugstores, among many other products.

Master Chao is the person in the foreground; in the background is Joe Perrott, Chumby's excellent project engineer from PCH China Solutions.

Master Chao is a craftsman in the traditional sense. It used to be that the finest furniture was designed and built only with the intuition and skill of a master craftsman. Now, we all go to IKEA and get CAD-designed, supply-chain-managed, picture-book-assembly furniture kits—and despite all that, it doesn't look too shabby. As a result, the word *craft* has been relegated to describe some scrapbook or needlepoint kit you buy at Michaels and put together on a slow weekend. We've forgotten that in an age before machines, "craft" was the only way anything of any quality was built.

It turns out, however, that traditional craft still matters, because CAD tools haven't brought about the ability to simulate our mistakes before we make them.

The creation of a *flat pattern* for textile goods is a good example of a process that requires a craftsman. A flat pattern is the set of 2D shapes used to guide the cutting of fabrics. These shapes are cut, folded, and sewn into a complex 3D shape. Mapping the projection of an arbitrary 3D shape onto a 2D surface with minimal waste area between the pieces is hard enough. The fact that the material stretches and distorts, sometimes in different directions, and that sewing requires ample tolerances for good yields, makes pattern creation a difficult problem to automate.

The chumby cases added another level of complexity, because they involved sewing a piece of leather onto a soft plastic frame. In that situation, as you sew the leather on, the frame distorts slightly and stretches the leather out, creating a sewing bias dependent upon the direction and rate of sewing. This force is captured in the seams and contributes to the final shape of the case. I challenge someone to make a computer simulation tool that can accurately capture those forces and predict how a product like that will look when sewn together.

Yet, somehow, Master Chao's proficiency in the art of pattern making enabled him to very quickly, and in very few iterations, create and tweak a pattern that compensated for all of those forces. His results, all obtained with cardboard, scissors, and pencils, were astoundingly clever and insightful. Be grateful for his old-world skills; they've likely played a role in the production of something you've used or benefited from.

There wasn't a single computer in Master Chao's office, yet the products I saw here wrapped around a wide array of high-tech devices.

Automation for Electronics Assembly

Before my work at Chumby, I thought almost everything was made by a machine. Of course, the tours of the textile factories corrected my impression very quickly; yet high-tech stuff like electronics assembly does still tend to be heavily automated, even in China. The only exceptions I saw during my factory tours were, ironically, the lowest-cost products, such as toys. These shops were still dominated by lines of workers, stuffing and dip-soldering circuit boards by hand.

One interesting dichotomy related to automation is the bimodal distribution of products that use *chip-on-board (CoB)* technology. CoB assembly directly bonds a silicon die to a PCB. Finished CoB assemblies have the distinctive "glob of epoxy" look to them, as opposed to the finished plastic-package look. High-end, dense electronics assemblies often employ CoB technologies. I've done a couple of CoB designs for some 10 Gb optical transceivers in my time, and they were not cheap.

At the same time, however, almost all toys use CoB technology, to eliminate the cost of the IC package! It's a testament to toy factories' tenacity about cost reduction that they would buy an automated wire bonder and stick it next to lines molding doll heads and sewing up stuffed animals because having an in-house wire bonder saves a nickel.

A typical wire bonder bonds a wire as thin as a human hair to a site on a silicon chip not much larger than the wire diameter, and it does this several times a second. Wire bonders are very fast, precise pieces of equipment. The bonding happens so quickly that the board seems to swivel smoothly around, but in fact, it stops 16 times as it spins around, and at each stop, a wire is bonded between the chip and the board.

Immediately before bonding, however, the chip is glued very carefully to the board by hand, and immediately after bonding, the chip is encapsulated by a human operator dispensing epoxy very carefully by hand. That means wire bonder is the only automated piece of equipment on assembly lines for simple toys. Seeing that process gave me a new appreciation for what goes into those talking Barney dolls that sell for $10 at Target.

The chumby manufacturing process used a bit of automation, too, courtesy of a chip shooter. Chip shooters (as well as pick-and-place machines) place surface-mount components on PCBs so the components can be soldered.

The chumby PCB assembly factory in China had dozens of lines filled with tried-and-true Fuji chip shooters.

It's absolutely mesmerizing to see a chip shooter in action. The chip shooters at the chumby PCB assembly factory were capable of placing 10,000 to 20,000 components per hour, per machine. This means that each machine could put down 3 to 6 components per second. The robotic assemblies move faster than the eye can see, and it all turns into an awe-inspiring blur. The chip shooter I saw at the chumby factory worked something like a Gatling gun: the chip gun itself was fixed, and the board danced around beneath the gun. The chip shooter actually "looked at" each component and rotated it to the correct orientation before putting it down on the board.

This is the end of the line for a chumby core board assembly!

The factory we used for the chumby's PCB assembly also produced name-brand PC motherboards and seemed to have no problem pushing out well over 10,000 such complex assemblies each day. But even though processes like component placement can be automated, there are some things a machine just can't do.

Precision, Injection Molding, and Patience

In the course of engineering the chumby, I also had to learn about injection molding, because the circuit board had to go inside a case of some kind. For an electronics guy with little mechanical background, this was no small hill to climb. The concept seems simple: you make a cavity out of steel, push molten plastic into it at high pressure, let it cool, and voilà—a finished part comes out, just like the Play-Doh molds from elementary school.

Oh, if only the process were that simple.

Sure, plastic flows, but it's not particularly runny. It moves slowly, and it cools as it flows. The color of the plastic is impacted by the temperature changes, and when using an improperly designed mold, you can even see flow lines and knit lines in the final product. There's also a whole assortment of issues with how the finished part is pulled from the mold, how the mold is made and finished, where the gates and runners are for getting the plastic inside the mold, and so on.

Fortunately, PCH had experts in China who knew all about this, and I got to learn mostly by watching.

If I were to summarize injection molding with a single adjective, it would be *precision*. When done right, the molds are precise to better than hair-thin tolerances, yet they are made out of hard steel. Achieving this level of precision out of such a durable material is no mean feat, and it's impressive to see a machine cut a mold out of raw steel.

The machine that cut the molds for the chumby case had a moving stage that rapidly pushed around a block of steel probably weighing several hundred pounds; it milled away at the metal in quite a hurry!

The mold-cutting machine used in manufacturing chumbys. Compare it to the people standing next to it for scale.

But machining is only the roughest step in mold making. After the rough shape is cut out, the mold is put into an *electrical discharge machine (EDM)*, where a burst of electrons knocks microscopic chunks off the steel surface. This is a terrifically tedious process: I've watched many EDMs do their job, and it's like watching paint dry. EDMs are, however, wicked precise, and they yield spectacular, repeatable results.

From a project management standpoint, the phenomenally long lead times of production-quality injection-molded plastics was the biggest eye opener for me. All told, the chumby mold transformed from a block of raw steel into a first-shot tool in four to six weeks, and I had to go to China and see the tooling shop do its work before I was convinced there wasn't some gross amount of schedule padding.

Even more harrowing from the risk management standpoint was the lack of good simulation tools to predict how plastics would flow through a mold. If we saw visible blemishes like flow lines and knit lines, we had to wait four to six weeks to see if the new mold was better. Ouch!

Fortunately, the toolmakers Chumby used in China anticipated these issues, and they made the tools to err on the side of excess steel, because removing material to fix a problem is much easier than adding material. It's like the old carpenter's saying: measure twice, cut once, and if you have to cut wrong, cut long.

The mold that was used to create the chumby's back bezel was extra complex, because it involved a process called *overmolding*. If you happen to own a chumby classic, look at the back side. There's a rubbery TPE surrounding the hard ABS bezel. Many people assumed this was a glued-on rubber band. In fact, the TPE is molded in place on the back piece. This requires a two-shot mold.

The final mold for the chumby's back bezel, inside an injection-molding machine

There were actually two molds, and one side of the mold spun around so that the alternating material systems could be molded at the right points in the process.

A lot of hard work goes into the humble plastic parts you see every day, and that's all part of creating quality products. But at the same time, there's also a very real need to meet the expectation of cheap prices.

The Challenge of Quality

Clearly, with the expectation of low cost of China-made goods comes a great challenge in quality management. Look at the media coverage on topics like lead paint in toys, industrial chemicals in food, and other items made in China, and you can see some of the bad decisions made to keep prices down.

When considering cases like that, I think it's important to apply Hanlon's razor. To paraphrase, "Never attribute to malice that which can be adequately explained by ignorance."

The Brits also have a nice, pithy version of the aphorism: "Cock-up before conspiracy."

Some manufacturers are indeed out there to make a buck at any cost, but I think the majority of mistakes are made out of ignorance. Most of the rank-and-file in factories don't know what their product is ultimately used for, and under intense pressure to reduce costs, they make those bad decisions. Factories also have to deal with products that are woefully underspecified, as well as customers who overwhelm them with all kinds of frivolous requirements—and most customers don't follow up in either case. In the end, the factories play a game of "ship and find out," and if the customer doesn't notice a missing spec, then the spec must not have been important. It's not a great game, and it means that customers need to be ever vigilant about audits and keeping the quality standard up.

THE DISCONNECT BETWEEN AMERICA AND CHINA

One fundamental problem behind this game is that many Chinese residents do not understand or appreciate basic things that we take for granted in America, and vice versa. Many Chinese factory workers are well educated, but they didn't grow up in a "gadget culture" like we have in the United States, so you can't assume anything about their abilities to subjectively interpret specifications for a product.

For example, you can tell a US engineer, "I'd like a button on that panel," and you'll probably get something pretty close to what you expect in terms of look and feel, since you and the engineer share common experiences and expectations for a button on a panel. If you did the same in China, you'd probably get something that looks a little awkward and has a clunky feel but is darn cheap and really easy to build and test. While the latter properties are desirable for practical reasons, American gadget connoisseurs just won't buy something that's aesthetically awkward or feels clunky.

Yet, ultimately, it's those consumers who want—nay, demand—low-priced goods, and that need drives the decision to manufacture in China. The trouble is that aside from the label on the product that says "Made in China" or "Made in the USA," consumers really don't care about the manufacturing process. What markup would you pay for a gadget that said "Made in the USA" on it? The cost premium for US labor is 10 times what it is in China. Think about it: can the average US factory worker be 10 times more productive than the average Chinese factory worker? It's a hard multiplier to play against.

I'm not saying there's no value in domestic vendors: it would be a lot less effort and less risk for me to get stuff made in the United States. In fact, most early prototypes are made there because of the enormous value that the domestic vendors can add. However, the pricing just doesn't work out for a mass-market product. Nobody would buy it, because its price wouldn't justify its feature set. One could even accuse me of being lazy if I were to just stick with a domestic vendor and pass the higher cost on to the customers.

BEING INVOLVED IN THE MANUFACTURING PROCESS

In the end, manufacturing in China is the best way to keep costs down, and to maintain quality, there is no substitute for going to China and getting directly involved. Almost every factory will "clean up" the day you come to visit, but with a sharp eye and the right questions, you can see through any quick veneers put in place.

When I evaluated factories for Chumby, I always visited the quality control (QC) room. I expected to see rows of well-maintained and well-worn binders with design documentation and QC standards, as well as *golden samples*, which are pre-production samples of a product. I'd demand to see the contents of a random binder and the golden sample associated with it, and verify that the employees knew what was going on in the binder. (Some factories do fill product binders with random data.) I also considered hard investments in equipment a good sign: the best manufacturers I visited all had a couple of rooms with sophisticated equipment for thermal, mechanical, and electrical limit testing, and of course, operators were in the room actually using the equipment. (I could definitely imagine a Chinese manufacturer buying a room of equipment just for show.)

But I suspect that toy manufacturers and food manufacturers don't fly technicians like me out to factories in China to oversee things on a regular basis. Contrast that with Apple, which regularly sends a cadre of engineers to work intense two-week (or longer) shifts in the factories (usually Foxconn, affectionately nicknamed "Mordor" by some at Apple). As a result, I bumped into many Apple engineers at the expat bars in Shenzhen.

The fact that PCH China Solutions offered Western-style management and quality control on site in China was important for us at Chumby. If we had a problem with a vendor, PCH sent someone to the factory right away to see what was going on—no phone tag, no FedEx filibuster. And factory owners in China tend to be very responsive when you show up at their doorstep.

Thus, Chumby's approach to the quality conundrum was holistic. We started by having an engineer (me) at the factory almost on day one to survey the situation. It's important to learn what the factory can and cannot do. I looked at what was being built on the line and what techniques were used. Then, when it came time to engineer the product, I tried to use the processes and techniques that were most comfortable for the factory. When I had to do something new (and any good, innovative product will need to), I picked my battles and focused on them, because anything new would be a multiweek challenge to get right. This strategy applies to even the smallest details: if the factory shrink-wraps goods in plastic, and you want to wrap your product in paper, then plan to focus heavily on developing the paper-wrapping process, because it's quite possible that none of the line workers at your factory of choice have even seen a paper-wrapped product before.

Of course, when developing a new process for the chumby, I preferred to be in the factory, and I still do. There's nothing like standing on the line and showing the workers who will be building your device how it should be made. For example, I personally trained the chumby assembly-line workers on how to attach a piece of copper tape to the LCD assembly to form a proper EMI shield.

It's difficult to describe the intricacies of how to fold tape across a complex piece of sheet metal to ensure it makes good electrical contact to the grounding surfaces without risking a short circuit to other components. Subtleties like the fact that the adhesive on one side is a poor insulator also require a basic understanding of physics that line workers simply don't have. Worse yet, explaining these concepts requires technical words that your translator might not even know.

In my case, even a good 3D drawing or photograph of the finished assembly couldn't have gotten the whole concept across, because the stiffness of the tape required a particular motion to fold without tearing. Describing the process remotely, approving samples via photographs, and ultimately approving a unit delivered via FedEx might have taken a couple of weeks, but standing in front of a group of workers and demonstrating the process firsthand took only a few minutes. And despite the language barrier, I could tell from their facial expressions and body language whether they understood the importance of a particular step. Given those cues, I immediately reviewed processes that were ambiguous or difficult to master.

Typically, when you can demonstrate a process at this level of detail and intimacy, the workers will get it right within hours, instead of weeks. This is part of the reason I spent so much time in China during the development of the chumby's manufacturing process.

Everyone was involved in the chumby quality process. This photo shows CEO Steve Tomlin (far left) and Artistic Director Susan Kare (middle) at the sewing factory, working out the details of logo silkscreening.

HOMEGROWN REMOTE TESTING

However, it wasn't always possible for Chumby to send someone to China. I, for one, preferred not to live in China, so at Chumby, we relied a lot on PCH to watch the quality and make sure things went well, and they did a superb job.

Often, working long distance meant that new processes took weeks to phase in if I wasn't there to tweak and approve on the spot, because every single tweak involved sending something almost round-trip through FedEx. After going through that process a few times, I learned to allocate two weeks per tweak, as opposed to the few hours it took when I was on the factory floor.

Those sets of two weeks added up fast.

Given the difficulty of overseeing operations in China from the United States, remote electronic monitoring of the products' test results was essential. For the chumby, I developed a set of testers that programmed, personalized, booted, verified, and measured every device off the assembly line. All data from the testing process was recorded to a log, and at the end of the day, the log was transferred to a server in the United States.

This data let me debug a plethora of problems on the floor. I could tell if an operator at a particular tester was having trouble with their barcode scanner. I also immediately knew if there was a yield problem that day, or if the throughput was slower than expected. It was very powerful to have this homegrown audit capability in place, because the factory knew I was watching them. In fact, having such a capability in place can make relationships with the factory run better: the factory eats the cost of yield problems (at least initially), so they appreciate it when the design engineer can offer expedient advice and help before any problems get out of hand.

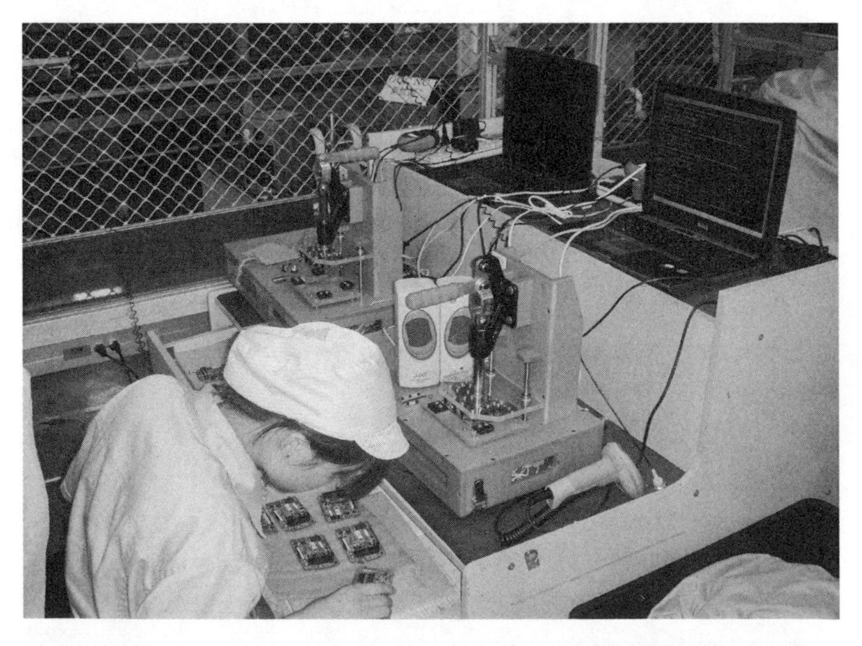

A pair of chumby test stations in the factory in China. There's quite a story about the trouble we went through getting those laptops into China.

FURTHER FACTORY TESTING

Once you've finished setting up the testing process, it can run autonomously at the factory. For example, at the chumby's PCB factory, the first pass of final inspection was done manually—one person went over every circuit board, and then with the help of a cardboard template, another operator ensured that no components were missing. The units then went on to automated testing.

Periodically, both PCH and the factory also performed Restriction of Hazardous Substances (RoHS) testing on chumby units to ensure that there was no contamination with a specified set of potentially harmful chemicals, including lead. RoHS is a hazardous chemical safety standard required in Europe but, ironically, not in the United States. Factories routinely do this test on all products, even those only shipping to the United States, because latent contamination on the line could prevent other products manufactured on the same line from shipping to Europe.

Even after all that testing, back in the United States, Chumby continued to sample units for QC purposes. To this end, we regularly ordered, characterized, and dissected devices to ensure that all the operating procedures were being followed.

MISTAKES STILL HAPPEN

Despite such safeguards, some mistakes will be made on any product. Every product goes through a phase where bugs that weren't caught by internal QA get pounded out. You have to rely on a top-notch customer service and support team, and you have to plan on being very agile and innovative during this phase to solve the problems and prevent them from ever happening again.

When I was at Chumby, if I heard about a unit in the wild with hardware problems, I actually called the customer who reported it. I wanted to know what went wrong so I could fix the problem and make sure it never happened again, to anyone!

My biggest hope with the chumby, however, was to avoid what happened to Microsoft and the Xbox 360's "red ring of death," where consoles would experience a major hardware failure, stop working, and just display a red light around the power button, causing huge frustration for players. This problem only exhibited itself after the Xbox 360 had been out for years, after millions of units had been shipped. Situations like the red ring of death are a product engineer's worst nightmare.

So you see, getting the chumby (or any product) to the point where it can ship to consumers is just the beginning. The real challenge starts after.

If you ever find yourself at this point in the manufacturing process, I wish you luck!

CLOSING THOUGHTS

The stories told here share some of my adventures—and failures—learning how to build products in volume. The next two chapters are more reflective and less narrative. The next chapter takes us on a virtual tour of three factories to see what we can learn from them, and Chapter 3 attempts to summarize all the lessons I've learned about manufacturing so far.

2. inside three very different factories

It's hard to understand how a computer works without opening it and looking around inside. Likewise, it's hard to understand how products are made without going into a factory and touring the line. Although we often think of manufacturing as the necessary but boring step after innovation, in reality, the two are tightly coupled. An inventor thinks about a product once; a factory thinks about the same product day in and day out, sometimes for years on end.

The importance of factories as an innovation node is only growing in today's connected global economy. The reality is that there is no "Apple factory" or "Nike factory." Rather, there is a series of facilities that are domain experts in processes (such as PCB fabrication or zipper manufacturing) that are curated by the familiar brands. Thus, it's not uncommon to see two competitors' products running side by side down similar lines in a single facility. This concentration of domain-specific expertise means that the best place to learn how to make an aspect of your product better is often the same place that makes a similar aspect in everybody else's products.

Some of the greatest insights I've had into improving a product have come from observing technicians at work on a line and seeing the clever optimization tricks they've developed after doing the same thing over and over for so long.

This chapter takes you on a tour of three factories that make everyday things: PCBs (in particular, the ones used in the Arduino), USB memory sticks, and zippers. By peeling back the curtain, you'll get some insight into the design trade-offs behind the products, and how they can be made better. In the PCB factory, I discovered the secret of how they print a high-resolution map of Italy on the back of every Arduino; in the USB memory stick factory, I witnessed a strange marriage of high- and low-tech manufacturing techniques; and in the zipper factory, I found out how even the humblest of products can bear valuable lessons for product designers.

WHERE ARDUINOS ARE BORN

It was July 2012, and it had been about six months since my previous startup, Chumby, ceased operations. I had decided to take a year off to figure things out and cross a few items off the bucket list, one of which was a trip to Italy. My girlfriend had the bright idea of reaching out to the Arduino team to see if I could visit their factory in Scarmagno (this was years before the Arduino/Genuino split) as part of our itinerary. Members of Officine Arduino (particularly managing director Davide Gomba) kindly took time out of their busy schedules to show me around their factory. They patiently waited as I expressed my inner shutterbug and general love for all things hardware, and I definitely came away with a lot of great photos.

A small town in northern Italy, Scarmagno is about an hour and a half west of Milan by car, near the Olivetti factories on the outskirts of Torino. The town handles all the circuit board fabrication, board stuffing, and distribution for officially branded Arduinos. I was really excited to see the factories, and the highlight of my tour was seeing System Elettronica, the PCB factory that made the Arduino PCBs.

One charming aspect of System Elettronica is that the owner painted the factory green, white, and red to match the colors of the Italian flag. On the factory floor, I saw some of that spirit in the red and green posts that ran the length of the facility.

A wide view of the factory floor at System Elettronica in August 2012

But I soon stopped paying much attention to the décor, as that factory floor was also where I got to follow a fresh batch of Arduino Leonardos through the entire manufacturing process. Here's how those boards were made.

Starting with a Sheet of Copper

Arduino Leonardo boards start as huge sheets of virgin copper-clad FR-4, a material made of fiberglass and epoxy that most PCBs use for a substrate, an insulating and structural layer between the copper layers. The sheets were 1.6mm thick (the most common thickness for a PCB, which corresponds to 1/16 inch), probably a meter wide, and about a meter and a half long.

A stack of copper sheets waiting to become Arduino boards

The first step in processing PCBs is to drill all the holes—pads, vias (the small holes that connect different layers of the PCB), mounting holes, plated slots, and so forth. When a PCB is manufactured, the holes are drilled before *patterning*, the stage where a masking chemical is photographically defined on the sheet everywhere the final boards need to have copper, including locations of traces, solder pads, and so on. Some of the drilled holes are used to align the masks that pattern the traces later in the process. Drilling is also a dirty and messy process that could damage circuit patterns if they were in place beforehand.

The CNC drilling head used to drill the Arduino boards

The blank copper panels were stacked three high, and a CNC drill took a single pass for all three, allowing it to drill three substrates at a time.

The drill rack used by the CNC drilling machine.
If you've ever had to create NC-drill files, this is that "drill rack."

Every hole in the Arduino board was mechanically drilled, including vias. The same is true of any PCB with through-holes, which is why the via count is such an important parameter in calculating the cost of a PCB.

Note that the particular drill I saw at System Elettronica was relatively small. I've seen massive drill decks in China that gang (mechanically attach) four or six drill heads together in a truck-size machine, processing dozens of panels at the same time as opposed to the three panels this drill could handle. The reasoning behind this approach is that the precise, robotic positioning assembly is the expensive part of a drilling machine. The drill itself is cheap—just a spinning motor to drive the bit. So, one way to increase throughput is to gang several drills together on one large assembly and move them in concert. Each individual drill still goes through its own stack of panels, but for the price of one X-Y positioner, you get four to six times the throughput as the drill I saw on my trip to Italy. Those bigger machines drill so fast and hard that the ground shakes with every via drilled, even from several meters away.

Once the panels are drilled, cleaned, and deburred, they are ready for the next step in the manufacturing process.

A stack of finished, drilled panels of Arduino Leonardo boards

Applying the PCB Pattern to the Copper

The next step is to apply a *photoresist*, a light-sensitive chemical, to the panel and expose a pattern. At System Elettronica, this process used a light box and a high-contrast film. I've also seen direct laser imaging—in the form of a raster-scanning laser—used to apply a pattern to a PCB. Direct laser scanners are more common in quick-turn prototype houses, and film imaging is more common in mass-production houses.

Before and after: the right panel shows photoresist prior to exposure, and the left panel after.

A PCB being mounted into a light box that will expose its unprocessed backside film

After the pattern is applied, each panel of boards is sent into a machine to be developed. In this case, the same machine is used to develop both the photoresist and the soldermask.

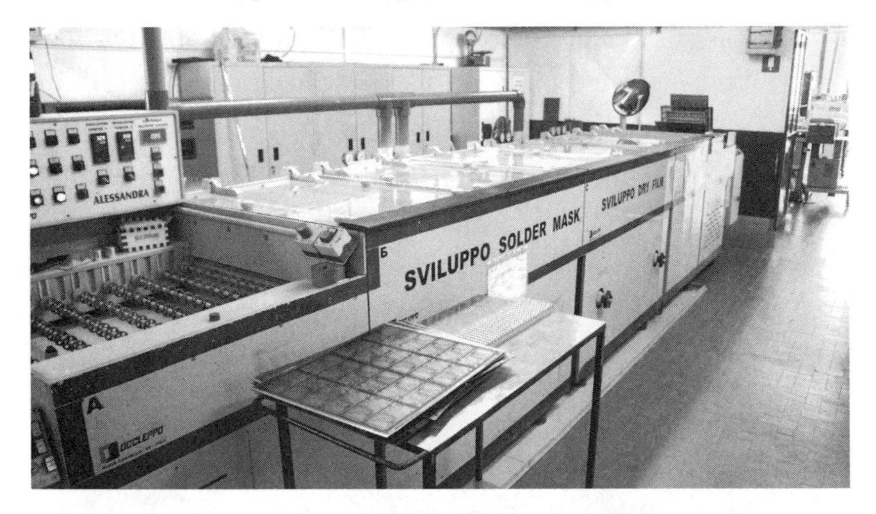

The machine that develops the photoresist

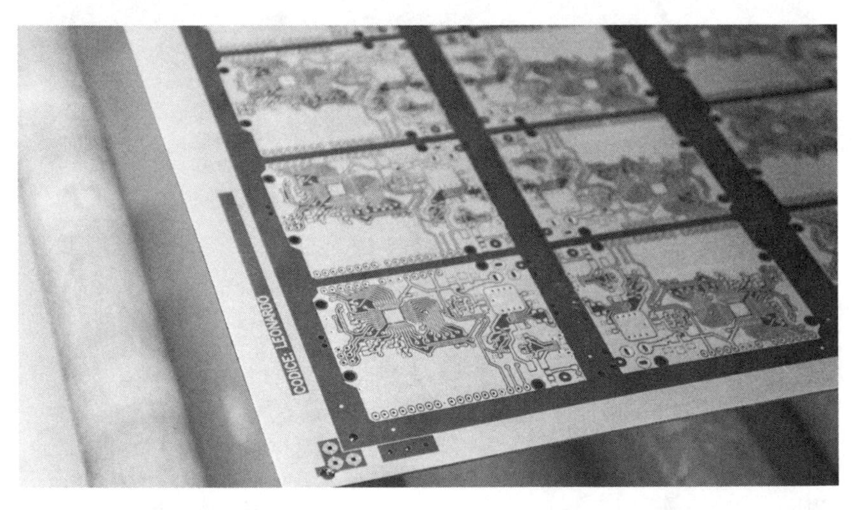

This photo of a panel with developed photoresist is one of my favorite photos from the System Elettronica factory. Also, something about "Codice: Leonardo" just sounds cool.

Etching the PCBs

After photo processing and development, the panels go through a series of chemical baths that etch and plate the copper.

The panels are swished gently back and forth in a chemical bath to expedite the etching process. The movement also circulates used etchant away from the panels, ensuring a more uniform etch rate regardless of the amount of copper to be removed. Moving the panels through these chemical baths was fully automated at Scarmagno. Automation is necessary because the panels must be treated with a series of caustic chemical baths with minimal exposure to oxygen. Oxygen can spoil a panel in a matter of seconds, so the transfer between the baths needs to be fast, and the amount of time a panel spends in a bath must be consistent. The baths also contain chemicals harmful to humans, so it's much safer for a robot to do this work.

A machine that moves panels around in etchant

Once the panels are processed in this series of solutions, a dull, white plating (which I'm guessing is nickel or tin) develops on all the surfaces of the panel not treated with photoresist, including the previously unplated through-hole vias and pads.

Panels of Arduino Leonardo boards after going through a series of chemical baths

At this point, the resist and unplated copper are stripped off, leaving just the raw FR-4 and the plated copper. The final step of processing produces a bright copper finish.

A panel etched of unwanted copper

PCB panels with bright, shiny copper. This photo doesn't show an Arduino panel, as those weren't going through the machine when I photographed it.

Applying Soldermask and Silkscreen

Once the copper is polished, the panels are ready for the *soldermask* (a protective, lacquer-like layer that insulates the copper traces below and prevents solder bridging above) and *silkscreen* (the ink used to label components, draw logos, and so on). These are applied in a process very similar to that of the trace patterns, using a photomask and developer/stripper machine.

A panel of Arduino boards with both soldermask and silkscreen developed

In the case of Arduinos, the silkscreen is actually a second layer of soldermask. A very specific formulation of dry-film white soldermask was procured for the Arduino team to create a sharp, good-looking layer that resolved the intricate artwork you see on Arduino boards—particularly the map of Italy on the backside. Other techniques I've seen for producing silkscreen layers include high-resolution inkjet printing, which is better suited for quick-turn board houses, and of course, the namesake squeegee-and-paint silkscreen process.

Testing and Finishing the Boards

After all that chemical processing, the panels receive a protective plating of solder from a hot-air solder leveling machine.

With the solder plating in place, every board is 100 percent tested. Every trace has its continuity and resistance measured with a pair of flying probes. The process I saw is called *flying head testing* (also referred to as *flying probe testing*), and in that sort of setup, several pairs of arms with needlelike probes test continuity between pairs of traces in a swift tapping motion. Considering all the traces on an Arduino Leonardo, that's a lot of probing! Fortunately the robot's arms move like a blur, as it can probe hundreds of points per minute.

NOTE *An alternative to flying head testing is clamshell testing, where a set of pogo pins is put into a fixture that can test the entire board with a single mechanical operation. However, clamshell fixtures are very labor-intensive to assemble and maintain, and require physical rewiring every time the Gerber files describing the PCB images are updated. So, in lower volumes, flying probe testing is more cost-effective and flexible than clamshell testing.*

A stack of near-finished PCB panels, ready for a final step of routing out the individual boards

This particular facility only created the panels; a different factory actually populated the components. In situations like that, before the panels can be sent to the next factory, the individual PCBs need to be routed so they'll fit inside *surface mount technology (SMT)* machines to have the components placed. The panels are once again stacked up and batch-processed through a machine that uses a router bit to cut and release the boards. After that, the boards are finally ready to ship to the SMT facility.

Several Arduino panels, stacked for routing

Smaller 2×6 panels make SMT processing more efficient.

A veritable stack of about 25,000 bare Arduino PCBs, ready to leave the PCB factory. From there, they were stuffed, shipped, and sold to makers around the world!

I'm glad I made the side trip to visit the Arduino PCB factory. I've visited several PCB factories, and every one has a different character and its own set of tricks to improve yield, as well as unique limitations that designers need to compensate for. It was also interesting to see the little trick about using an extra layer of soldermask instead of silkscreen for achieving high cosmetic quality. While the resolution of a silkscreen is limited by the mesh of the silk barrier to hold the paint, soldermask is limited by the quality of the optics and chemical developing, giving over an order of magnitude improvement in resolution and ultimately a higher perceived quality. Normally the lower quality of silkscreen is acceptable because end users don't see the circuit boards inside computers, but for Arduino, the end product *is* the circuit board.

WHERE USB MEMORY STICKS ARE BORN

Several months after my tour of the Arduino factory, I had the good fortune of being a keynote speaker at Linux Conference Australia (LCA) 2013. In my talk, "Linux in the Flesh: Adventures Embedding Linux in Hardware," I discussed how Linux is in all kinds of devices we see every day. This story isn't about Linux, but it does connect me and, tangentially, LCA to a factory.

One of the tchotchkes I received from the LCA organizers was a little USB memory stick with Tux the penguin, the Linux mascot, on the outside. When I saw the device, I thought it was a neat coincidence that about a week before the conference, I had been in a factory that manufactured USB memory sticks exactly like it. I saw the USB stick board assembly process from start to finish, and it surprisingly involved a lot less automation than the Arduino manufacturing process did.

The Beginning of a USB Stick

USB sticks start life as bare flash memory chips. Prior to being mounted on PCBs, these chips are screened for memory capacity and functionality.

A workstation where flash memory chips are screened.
The metal rectangle on the left with the circular cutaway is the probe card.

At a workstation in this factory, stacks of bare-die flash chips awaited testing and binning with a *probe card*, which has tiny, very accurately positioned pins used to touch down on pads only a little bit wider than a human hair on a silicon wafer's surface. (I love how the worker at this particular station used rubber bands to hold an analog current meter to the probe card.)

The probe card, up close

Looking through the microscope on the microprobing station. Notice the needles touching the square pads at the edge of the flash chip's surface. Each pad is perhaps 100 microns on a side—a human hair is about 70 microns in diameter.

Interestingly, the chips I saw were absolutely not tested in a clean-room environment. Workers handled chips with tweezers and hand suction vises and mounted the probe cards into their jigs by hand.

Hand-Placing Chips on a PCB

Once the chips were screened for functionality, they were placed *by hand* onto the USB stick PCBs. This is not an unusual practice; every value-oriented wire-bonding facility I've visited relies on the manual placement of bare die.

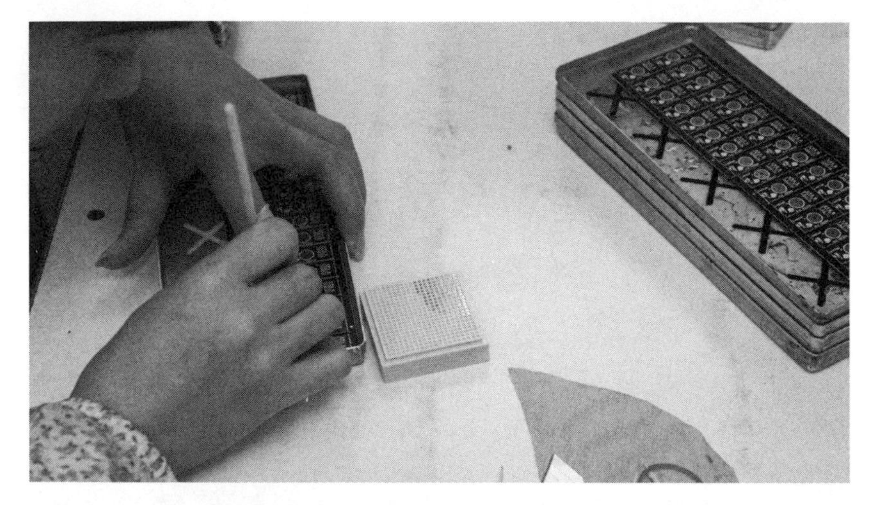

A controller IC being placed on a panel of USB-stick PCBs. The tiny bare dies are on the right, sitting in a waffle pack.

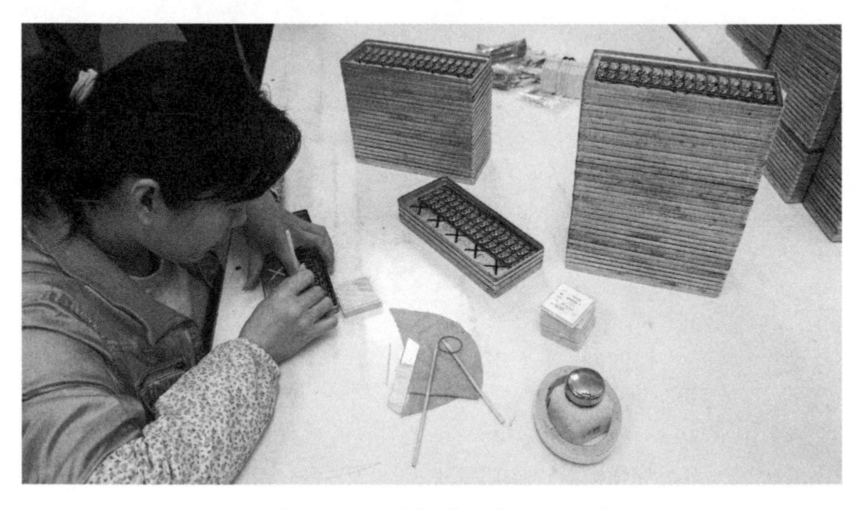

A zoomed-out view of the die-placing workstation

The lady I watched placing the bare die was using a chopstick-like tool made of hand-cut bamboo. I still haven't figured out exactly how the process works, but my best guess is that the bamboo sticks have just the right surface energy to adhere to the silicon die, such that silicon sticks to the tip of the bamboo rod. A dot of glue is preapplied to the bare boards, so when the operator touches the die down onto the glue, the surface tension of the glue pulls the die off of the bamboo stick.

It's trippy to think that the chips inside my USB stick were handled using modified chopsticks.

Bonding the Chips to the PCB

Once the chips were placed on the PCB, they were *wire bonded* to the board with an automated bonding machine, which uses computer-assisted image recognition to find the location of the bond pads (this is part of the reason the factories can get away with manual die placement). Wire bonding is the process that connects an integrated circuit to its packaging, and the automated bonding machine connected wires to the IC at an insane speed, rotating the circuit board all the while. As I watched this process, the operator had to pull off and replace a misbonded wire by hand and then refeed the wire into the machine. Given that these wires are thinner than a strand of hair and that the bonding pads on the packaging and the IC are microscopic, that was no mean feat of manual dexterity.

A Close Look at the USB Stick Boards

Just as the Arduino factory used panels containing multiple Leonardo boards, the USB memory stick factory used panels of eight USB sticks each. Each stick in the panel consisted of a flash memory chip and a controller IC that handled the bridging between USB and raw flash, a nontrivial task that includes managing bad block maps and error correction, among other things. The controller was probably an 8051-class CPU running at a few dozen MHz.

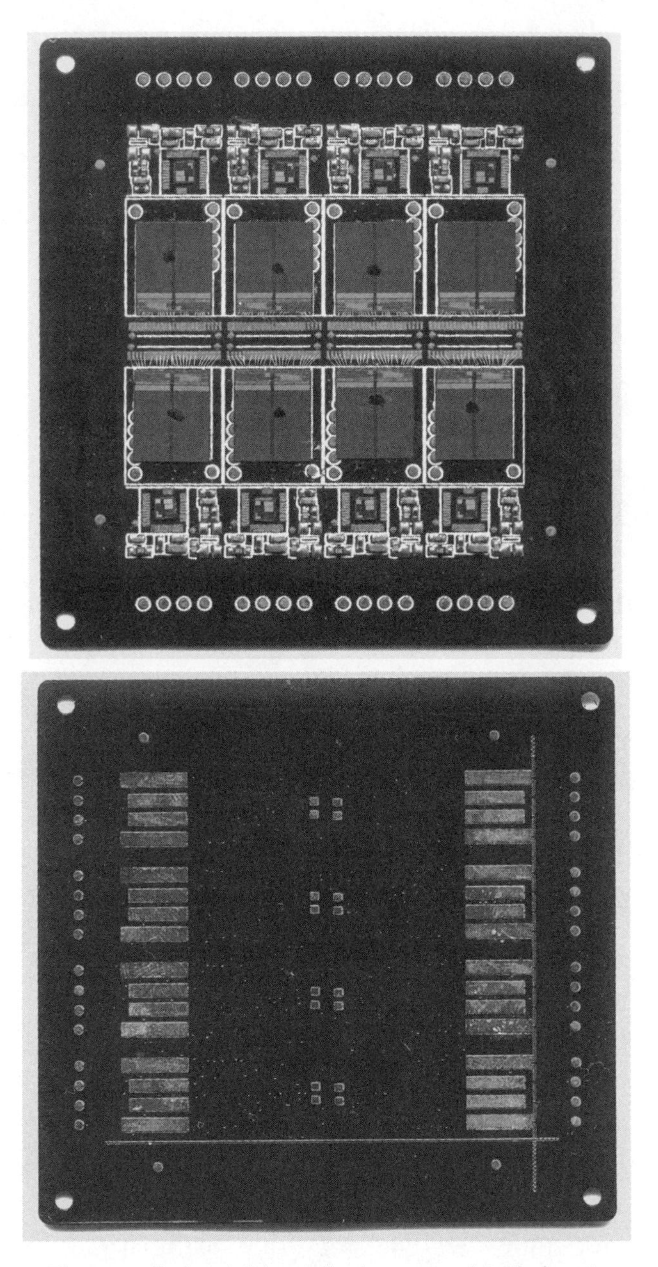

The partially bonded but fully die-mounted PCB that the factory owner gave me as a memento from my visit. Some of the wire bonds were crushed in transit.

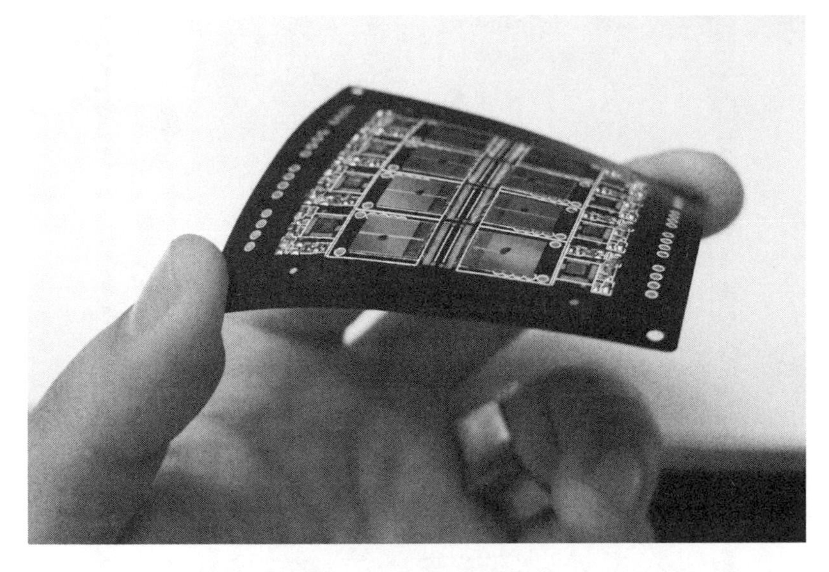

Interestingly, the entire USB stick assembly is flexible prior to encapsulation.

The die marking from the flash chip. Apparently, it's made by Intel.

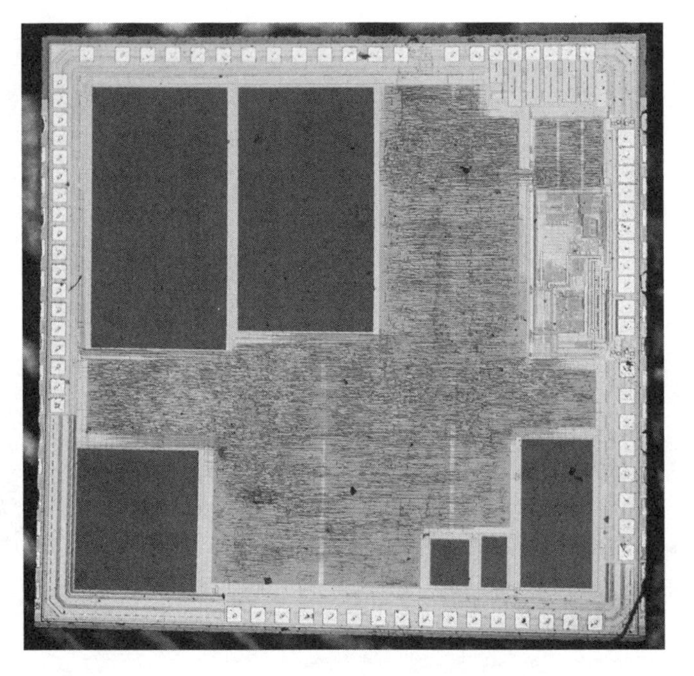

A die shot of the controller chip that went inside the USB sticks

Once the panels were bonded and tested, they were overmolded with epoxy and then cut into individual pieces, ready for sale.

But that's enough about electronics manufacturing; next, I want to show you a different kind of factory floor.

A TALE OF TWO ZIPPERS

My friend Chris "Akiba" Wang has a similar background to mine, except in his younger years he was way hipper: he was a dancer for acts like LL Cool J and Run DMC in the '90s. After going through a phase working for big semiconductor companies, he eventually quit and followed his passion to design and manufacture his own hardware projects. An expert in short-range, low-power wireless networking (he's co-authored a book on Bluetooth low energy and sells an Arduino + 802.15.4 variant called the "Freakduino"), he now consults for organizations like the United Nations and Keio University, runs FreakLabs, and collaborates with various dance acts, such as the Wrecking Crew, to provide unique and compelling lighting solutions for stage shows.

I had the good fortune of introducing Akiba to the greater Shenzhen area on a trip with MIT Media Lab students in 2013—the same trip where we toured the USB memory stick factory. Since then, he's been exploring deeper and deeper into the area. As his work spans the disciplines of performance art, wearables, and electronics, his network of factories is quite different from mine, so I always relish the opportunity to learn more about his world.

In January 2015, Akiba took me to visit his friend's zipper factory. I was very excited for the tour: no matter how humble the product, I always learn something new by visiting its factory. This factory was very different from both the Arduino and the USB stick facilities. There were far fewer employees, and it was a highly automated, vertically integrated manufacturer. To give you an idea of what that means, this facility turned metal ingots, sawdust, and rice into zipper parts.

Approximately 1 ton of ingots,
composed of 93 percent zinc and 7 percent aluminum alloy

Compressed sawdust pellets, used to fuel the ingot smelter

Rice, used to feed the workers

Finished zipper puller and slider assemblies

Let's look at one side of how that process actually works.

A Fully Automated Process

Between the three input materials and the output product was a fully automated die-casting line to create the zipper pullers and sliders, a set of tumblers and vibrating pots (or, as I like to call them, "vibrapots") to release and polish the zippers, and a set of machines to deburr and join each puller to its slider. I think I counted fewer than a dozen employees in the facility, and I'm guessing their capacity well exceeds a million zippers a month.

I was mesmerized by the vibrapots* that put the zippers together. There were two vibrapots: one with pullers and one with sliders. Both sliders and pullers were deposited onto a moving rail, and as I watched these miracles at work, it looked as if the sliders and pullers were lining themselves up in the right orientation by magic. Each fell into its rail, and at the end of the line, they were pressed together into a familiar zipper form, all in a single, fully automated machine.

* I honestly don't what they're called, so yes, I'm going to keep calling them that.

When I put my hand in the pot, I found there was no stirrer to cause the motion; I just felt a strong vibration. I relaxed my hand, and found it started to move along with all the other items in the pot. The entire pot was vibrating in a biased fashion, such that the items inside tended to move in a circular motion. This pushed the pullers and sliders onto the set of rails, which were shaped to take advantage of asymmetries in the objects to allow only the pieces that jumped on the rail in the correct orientation to continue to the next stage.

A Semiautomated Process

Despite the high level of automation in this factory, many of the workers I saw were performing one operation. They fed the pullers for a different kind of zipper into a device connected to another vibrapot containing sliders, while the device put the sliders and pullers together.

Of course, I asked, "Why do some zippers have fully automated assembly processes, whereas others are semiautomatic?"

The answer, it turns out, is very subtle, and it boils down to shape.

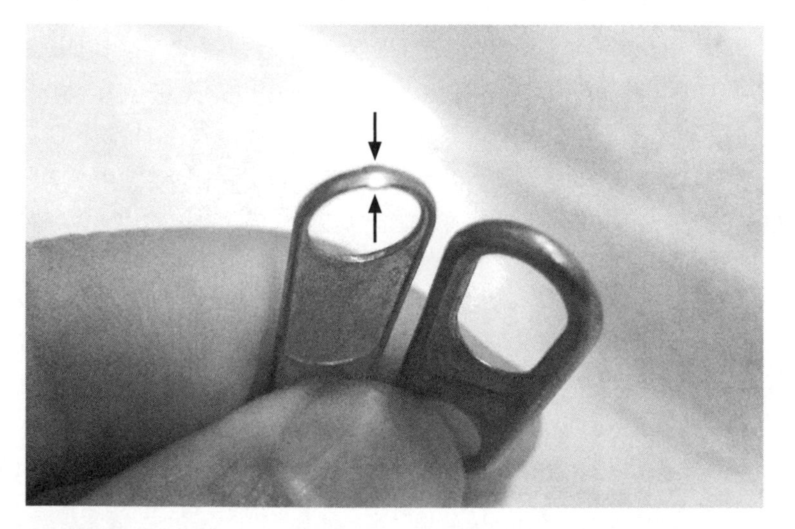

Note the difference in these two pullers, indicated by the arrows.

One tiny tab, barely visible, was the difference between full automation and needing a human to join millions of sliders and pullers together. To understand why, let's review one critical step in the vibrapot operation. A worker kindly paused the vibrapot responsible for sorting the pullers into the correct orientation for the fully automatic process so I could take a photo of the key step.

Pullers coming through the vibrapot

When the pullers came around the rail, their orientation was random: some faced right, some left. But the joining operation must only insert the slider into the smaller of the two holes. That tiny tab allowed gravity to cause all the pullers to hang in the same direction as they fell into a rail toward the left.

The semiautomated zipper design doesn't have this tab; as a result, the design is too symmetric for a vibrapot to align the puller. I asked the factory owner if adding the tiny tab would save this labor, and he said absolutely.

At this point, it seemed blindingly obvious to me that all zippers should have this tiny tab, but the zipper's designer wouldn't have it. Even though such a tab is very small, consumers can feel the subtle bumps, and some perceive it as a defect in the design. As a result, the designer insisted upon a perfectly smooth tab, which accordingly had no feature to easily and reliably allow for automatic orientation.

The Irony of Scarcity and Demand

I'd like to imagine that most people, after watching a person join pullers to sliders for a couple of minutes, would be quite content to suffer a tiny bump on the tip of their zipper to save another human the fate of manually aligning pullers into sliders for eight hours a day. Alternatively, I suppose an engineer could spend countless hours trying to design a more complex method for aligning the pullers and sliders, but there are two problems with that:

- The zipper's customer probably wouldn't pay for that effort.
- It's probably net cheaper to pay unskilled labor to manually perform the sorting.

This zipper factory owner had already automated everything else in the facility, so I figure they've thought long and hard about this problem, too. My guess is that robots are expensive to build and maintain; people are self-replicating and largely self-maintaining. Remember that third input to the factory—rice? Any robot's spare parts have to be cheaper than rice for the robot to earn a place on this factory's floor.

In reality, however, it's too much effort to explain this concept to end customers; in fact, quite the opposite happens in the market. Putting the smooth zippers together involves extra labor, so the zippers cost more; therefore, they tend to end up in high-end products. This further enforces the notion that really smooth zippers with no tiny tab on them must be the result of quality control and attention to detail.

My world is full of small frustrations like this. For example, most customers perceive plastics with a mirror finish to be of a higher quality than those with a satin finish. There is no functional difference between the two plastics' structural performance, but making something with a mirror finish takes a lot more effort. The injection-molding tools must be painstakingly and meticulously polished, and at every step in the factory, workers must wear white gloves. Mountains of plastic are scrapped for hairline defects, and extra films of plastic are placed over mirror surfaces to protect them during shipping.

For all that effort, for all that waste, what's the first thing users do? They put their dirty fingerprints all over the mirror finish. Within a minute of a product coming out of the box, all that effort is undone. Or worse yet, the user leaves the protective film on, resulting in a net worse cosmetic effect than a satin finish.

Contrast this to satin-finished plastic. Satin finishes don't require protective films, are easier for workers and users to handle, last longer, and have much better yields. In the user's hands, they hide small scratches, fingerprints, and bits of dust. Arguably, the satin finish offers a better long-term customer experience than the mirror finish.

But that mirror finish sure does look pretty in photographs and showroom displays!

3. the factory floor

The previous two chapters were filled with stories of my personal experiences learning, making mistakes, and growing with the manufacturing ecosystem in the greater Shenzhen area. In January 2013, after I'd learned the ropes, the MIT Media Lab asked me to start mentoring graduate students on supply chain and manufacturing, and I took them on a tour of Shenzhen (the same tour where I met Akiba and visited the USB memory stick factory). This chapter is an attempt to distill everything I taught over a course of weeks into a couple dozen pages.

The challenges and trade-offs in low-volume manufacturing are different from those of well-funded corporate exercises that prototype at the scale of thousands of units. I learned this over time, but not everyone has six years to bumble through all the newbie mistakes. If you're already in a fast-moving tech startup, you probably don't have the luxury of doing any exploration at all. The lessons in this chapter are applicable to anyone looking to bootstrap a hardware product from an initial prototype to moderate volumes (perhaps hundreds of thousands of units). Treat this summary as a general guideline, not a detailed roadmap. The devil is always in the details, and one fun part of making new, innovative hardware products is there's no end of novel and interesting challenges to be solved.

HOW TO MAKE A BILL OF MATERIALS

Most makers trying to scale up their output quickly realize the only practical path forward is to outsource production. If only outsourcing were as easy as schematic + cash = product!

Whether you work with the assembly shop down the street or send your work to China, a clear and complete *bill of materials (BOM)* is the first step to outsourcing production. Every single assumption you make about your circuit board, down to the color of the soldermask, has to be spelled out unambiguously for a third party to faithfully reproduce your design. Missing or incomplete documentation is the leading cause of production delays, defects, and cost overruns.

A Simple BOM for a Bicycle Safety Light

For a case study, suppose you ran a successful Kickstarter campaign for a bicycle safety light. It contains a circuit that uses a 555 timer to flash a small array of LEDs. After a great marketing campaign, several hundred orders need to be filled in a few months' time.

At first, a BOM for the bicycle light, as automatically generated by a design tool such as Altium, might look like this:

Quantity	Comment	Designator
1	0.1μF	C1
1	10μF	C2
3	white LED	D1, D2, D3
1	2N3904	Q1
1	100	R1
2	20k	R2, R4
1	1k	R3
1	555 timer	U1

A very basic bicycle safety light BOM

This BOM, along with a schematic, is likely sufficient for any graduate of a US electrical engineering program to reproduce the prototype, but it's far from adequate for a manufacturing cost quotation. This version of the BOM addresses only electronics. A complete BOM for an LED flasher also needs to include the PCB, battery, plastic case pieces, lens, screws, any labeling (like a serial number), a manual, and packaging (plastic bag plus cardboard box, for example). It may also need a master carton to ship multiple LED flashers together, as a single boxed LED flasher is too small to ship on its own. Although cardboard boxes are cheap, they aren't free, and if they aren't ordered on time, inventory will sit on the dock until a master carton is delivered for final pack-out prior to shipment.

The following key information is also missing:

- Approved manufacturer for each component
- Tolerance, material composition, and voltage specification for passive components
- Package type information for all parts
- Extended part numbers specific to each manufacturer

Let's look at each of the missing items in more detail.

Approved Manufacturers

A proper factory will require you to supply an *approved vendor list (AVL)* specifying the allowed manufacturer(s) for every part on a PCB. A manufacturer is not a distributor but rather the company that actually makes a part. A capacitor, for example, could be made by TDK, Murata, Taiyo Yuden, AVX, Panasonic, Samsung, and so on. I'm still surprised at how many BOMs I've reviewed list DigiKey, Mouser, Avnet, or some other distributor as the manufacturer for a part.

It may seem silly to trifle over who makes a capacitor, but there are definitely situations where the maker of a component matters—even for the humble capacitor. For example, blindly substituting the filter capacitors on a switching regulator, even if the substitute has the same rated capacitance and voltage, can lead to unstable operation and even boards catching fire.

Of course, some parts in a design can be truly insensitive to the manufacturer, in which case I would mark "any/open" on the BOM for the AVL. (This is particularly true for parts like pull-up resistors.) This invites the factory to suggest their preferred supplier on your behalf.

Tolerance, Composition, and Voltage Specification

For passive components marked "any/open," you should always specify the following key parameters to ensure the right part is purchased:

- For resistors, specify at minimum the tolerance and wattage. A 1 kΩ, 1 percent tolerance, 1/4 W carbon resistor is a very different beast from a 1 kΩ, 5 percent tolerance, 1 W wire-wound resistor!
- For capacitors, specify at minimum the tolerance, voltage rating, and dielectric type. For special applications, also specify certain parameters such as ESR or ripple current

tolerance. A 10 μF, electrolytic, 10 percent tolerance capacitor rated for 50V has vastly different performance at high frequencies compared to a 10 μF, ceramic, 20 percent tolerance capacitor rated for 16V.

Inductors are sufficiently specialized that I don't recommend ever labeling them as "any/open" in your BOM. For power inductors, the basic parameters to specify are core composition, DC resistance, saturation, temperature rise, and current, but unlike resistors and capacitors, inductors have no standard for casing. Furthermore, important parameters such as shielding and potting, which can have material impacts on a circuit's performance, are often implicit in a part number; hence, it's best to fully specify the inductor. The same goes for RF inductors.

Electronic Component Form Factor

Always fully specify the *form factor*, or package type, of a component. Poorly specified or underspecified package parameters can lead to assembly errors. Beyond basic parameters like the Electronic Industries Alliance (EIA) or JEDEC Solid State Technology Association package code (that is, 0402, 0805, TSSOP, and so on), consider the following package information as you create your BOM:

Surface mount packages The height of a component can vary, particularly for packages larger than 1206 or for inductors. Pay attention to whether the board is slotting into a tight case.

Through-hole packages Always specify lead pitch and component height.

For ICs in general, try to also specify the common name that corresponds to the package, not just the manufacturer's internal code. For example, a Texas Instruments "DW" type package code corresponds to an SOIC package. This consistency check helps guard against errors.

Extended Part Numbers

Designers often think about components in abbreviated part numbers. A great example of this is the 7404. The venerable 7404 is a hex inverter and has been in service for decades. Because of its ubiquity, *7404* can be used as a generic term for an inverter among design engineers.

When going to production, however, you must specify information like the package type, manufacturer, and logic family. A complete part number for a particular hex inverter might be **74**VHCT**04**AMTC, which specifies an inverter made by Fairchild Semiconductor, from the VHCT series, in a TSSOP package, shipped in tubes. The extra characters are very important, because small variations can cause big problems, such as quoting and ordering the wrong packaged device and being stuck with a reel of unusable parts or subtle reliability problems.

For example, on a robotics controller I designed (codenamed *Kovan*), I encountered a problem due to a mistaken substitution of *VHC* in the part number for a component in the *VHCT* logic family. Using the VHC part switched the input thresholds of the inverter from TTL to CMOS logic-compatible, and some units had an asymmetric response to input signals as a result. Fortunately, I caught this problem before production ramped. The correct part was used on all other units, and I avoided a whole lot of potential rework—or worse, returns from upset customers. Luckily, the only cost of the mistake was reworking the few prototypes I was validating before production.

Here's another example of how missing a few characters in a part number can cost thousands of dollars. A fully specified part number for the LM3670 switching regulator might be LM3670MFX-3.3/NOPB. If */NOPB* is omitted, the part number is still valid and orderable—but that version uses leaded solder. This could be disastrous for products exporting to a region that requires RoHS compliance (meaning lead-free, among other things), like the European Union.

The *X* in the part number is another, more subtle issue. Part numbers with an *X* come in reels of 3,000 pieces, and those lacking an *X* come in reels of 1,000 pieces. While many factories will question an */NOPB* omission since they typically assemble RoHS documentation as they purchase parts, they rarely flag the reel quantity as an issue.

But *you* should care about the reel quantity. If you plan to build only 1,000 products, including the *X* in the part number means you'll have 2,000 extra LM3670s. And yes, you're on the hook to pay for the excess, since your BOM specified that part number. There are many valid reasons for ordering excess parts, so factories will rarely question a decision like that.

On the other hand, parts ordered in lots of 1,000 units are a bit more expensive per unit than those ordered in lots of 3,000. So, if you leave out the *X* as your volume increases, you'll end up paying more for the part than you have to. Either way, the factory will quote your BOM exactly as specified, and if your quantity specifiers are incorrect, you could be leaving money on the table—or worse, losing money.

The bottom line? Every digit and character counts, and lack of attention to detail can cost real money!

The Bicycle Safety Light BOM Revisited

With those four points in mind, consider how a proper, fully specified BOM for the bicycle safety light example might look.

Qty	Value	Package	Designator	AVL1	AVL1 P/N	MOQ	Lead time
1	0.1µF, ceramic, 25V, 10%, X5R	0402	C1	Taiyo Yuden	TMK105BJ104KV-F....	10000	8 wks
1	10µF, ceramic, 16V, 10%, X5R	1206	C2	TDK	C3216X5R1C106K(085AB)	2000	12 wks
3	white LED, water clear lens	T-1 ¾	D1, D2, D3	Lumex	SSL-LX5093UWC/G	3000	12 wks
1	2N3904	SOT-223	Q1	ON Semiconductor	PZT3904T1GOS	1000	6 wks
1	100 ohm, 1/2W, 5%	2010	R1	Panasonic	ERJ-12SF100U	5000	8 wks
2	20k, 1/16W, 1%	0402	R2, R4	any/open		10000	8 wks
1	1k, 1/16W, 5%	0402	R3	any/open		10000	8 wks
1	NE555D	SOIC-8	U1	TI	NE555D	1000	4 wks
1	PCB, FR4, 1.6mm +/- 10%, green soldermask, HASL, white silkscreen, 5cm × 8cm		PCB	TBD	FLASHYLIGHT_GERBERS_V1.ZIP	1000	4 wks
1	Plastic ABS, bottom case, satin finish, lead free, black			TBD	FLASHYLIGHT_BOT_V1.STEP	1000	16 wks / 4 wks
1	Plastic ABS, top case, satin finish, lead free, black			TBD	FLASHYLIGHT_TOP_V1.STEP	1000	16 wks / 4 wks
1	Plastic polycarbonate, lens, mirror finish, lead free, clear			TBD	FLASHYLIGHT_LENS_V1.STEP	1000	16 wks / 4 wks
4	Screw, M2x4, pan head philips, self-tapping 5mm			any/open		4000	stock
1	Battery Snap, 9V, 15CM red and black 26 AWG wires (5mm Leads)			Kaweei	CBS-150	5000	1 wk
1	Instruction manual, A4 sheet, black and white, two sides printed			any/open	flashylite_manual_v2.ai	1000	3 wks
1	10cm × 12cm PE plastic bag, clear			any/open		1000	1 wk
1	Bar code label, serial number and date code, CODE39 5mm × 15mm			any/open	barcode_sample_v1.pdf	1000	1 wk
1	Cardboard box, 6cm × 6cm × 10cm, natural color, 50lb stock			any/open	see included box sample	1000	1 wk
0.02	Master carton, 60cm × 40cm × 20cm			any/open		100	1 wk

The improved bicycle safety light BOM

There's a big difference between a BOM that any engineer could use to produce a prototype, like the first one I showed for the bicycle safety light, and a BOM like this, which any factory could use to mass-produce a product. Notice the MOQ (minimum order quantity) and Lead Time columns in particular. These columns are irrelevant when you're building low-volume prototypes, as you'd typically buy parts from distributors that have few MOQ restrictions and maintain stock for next-day deliveries. When scaling into production, however, you save a lot of money by cutting the distributor overhead and buying through wholesale channels. In wholesale channels, MOQs and lead times matter.

The good news is that the factory will fill in the MOQ and lead time as part of the quotation process. But you'll find it helpful to track these parameters from the beginning. If the MOQ of a particular component is very high, the factory may have to buy massive numbers of excess parts, which increases the effective price of the project. If the lead time of a part is very long, you may want to consider redesigning for a part with a shorter lead time. Using parts with shorter lead times not only saves time but also improves cash flow: no one wants to tie up cash on long-lead components four months in advance of sales revenue.

This BOM also includes several nonelectronic items—like the box, a bar code label, and so on—which wouldn't be on the engineering prototype's BOM. These miscellaneous bits are easy to forget, but a missing user manual in an initial BOM is often not discovered until the final sample is opened for approval, leading to a last-minute scramble to get the manual into the final product. Many products have been delayed simply because a user manual or box art wasn't completed and approved in time, and it sucks to have a hundred thousand dollars' worth of inventory idling in a warehouse for want of a slip of paper.

Beyond a proper BOM, providing the factory with golden samples of your product along with your CAD files is another best practice. These working prototypes enable the factory to make smarter decisions about any ambiguities in your submitted BOM. Hand-soldering one more unit just for the factory may seem annoying, but in my opinion, a few hours of soldering beats a week of trading emails with the factory.

NOTE *When you're building a business model, parts and packaging still aren't the only costs to consider. Even this detailed BOM doesn't list factory margin, labor for assembly, packout, shipping, duties, and so on. I discuss these "soft costs" in "Picking (and Maintaining) a Partner" on page 107.*

Planning for and Coping with Change

Of course, even if your design is perfect and your BOM is ideal, your design may still have to change if vendors *end-of-life (EOL)*, or stop making, components you selected. And let's face it: there's always a chance your design assumptions won't survive contact with real consumers, too.

Before crossing the threshold into production, formalize the process for changing a design with the factory. It's best practice to use written, formal *engineering change orders (ECO)* to update the factory on any changes after the initial quotation. At minimum, here's what an ECO template should include:

- The details of each changed part, and a brief explanation of why the change is needed
- A unique revision number for conveniently referencing the change down the road
- A method to record the factory's receipt of the ECO paperwork

Be thorough with ECOs, rather than relying on casual emails, or the buyers at your factory may buy the wrong part. Worse yet, the factory might *install* the wrong part, and entire lots of your product will need to be scrapped or reworked. Even after troubleshooting a problem with the factory engineers, I still write up a formal ECO and submit it to the production staff to formalize the findings. I hate paperwork as much as the next engineer, but in production, one small mistake can cost tens of thousands of dollars, and that thought keeps me disciplined on ECOs.

On the next page is an actual ECO I issued that ended up saving me time and money.

Note the date on this ECO: February 27, 2014. This ECO was issued right before the Chinese New Year, when the factories go on holiday for a couple of weeks. There is significant turnover of unskilled labor inside factories after the holidays, and thus there's a lot of opportunity for work orders to get lost and forgotten. Worried that the ECO would be missed, I consulted with the managers after the factory resumed production to ensure the ECO wasn't forgotten. They assured me it was applied, but I still felt a vague paranoia, so I asked for photos of the circuit board to confirm. Sure enough, the first production batch was missing the change in my ECO.

Thanks to the detailed ECO, the factory readily admitted its error, repaired the entire production run, and paid for the reworking. But if I'd sent the change order in a quick email without referencing specific batches or work orders, there could have been sufficient ambiguity for the factory to get out of the rework charges. The factory could have argued that it thought I meant to apply the change to a future production run, or it could simply deny receiving a confirmed order, as emails are a fairly casual form of communication. Either way, a few minutes of documentation saved days of negotiation and hundreds of dollars in rework fees.

ENGINEERING CHANGE ORDER

Date: 27 February 2014
Sutajio Ko-Usagi Pte LTD
bunnie@.com

ECO number: 0001 version: 2
Project:
Subassembly: sensor and microcontroller
Reference PO: PO-0018 and PO-0016

Background

Per request by engineer , pull-ups on inputs to the microcontroller and trigger sticker are to be modified to enhance flexibility and better target user use-cases.

On the microcontroller, R2, R3, and R4 (all 22M, 5%) shall be omitted, to allow the inputs to be used in applications that bar the presence of a pull-up.

On the trigger, R16 shall be changed from 10k, 1% to 22M, 5%, to allow for resistive-touch style sensing of the input pin.

CHANGE ORDER DETAILS

ORIGINAL		NEW		Comments
Designator	Value	Designator	Value	
R2	22M, 5% 0603	R2	DNP	BOM change only
R3	22M, 5% 0603	R3	DNP	BOM change only
R4	22M, 5% 0603	R4	DNP	BOM change only
R16	10k, 1% 0603	R16	22M, 5% 0603	BOM change only

MATERIAL DISPOSITION

No extra material needs to be ordered to execute this change.

Excess material resulting from this change shall be held by and applied to future builds. No expected change to PO or cost for assembly.

Version history

version 2 – changed 0805 to 0603 for part footprints, was a typo.

Example of an actual ECO used in production. Thanks to the formal documentation process, a production mix-up related to this ECO was resolved in my favor.

PROCESS OPTIMIZATION: DESIGN FOR MANUFACTURING

While you're designing your final product and putting together a BOM, considering *yield*, the number of good units that come out of the manufacturing process, is also important. Yield is a boring subject for many engineers, but for entrepreneurs, success or failure will be determined in part by whether they achieve a reasonable yield. Fortunately, you can help your yield by designing with it in mind.

Why DFM?

Unlike software, every copy of a physical good has slight imperfections. Sometimes the imperfections cancel out; sometimes, they gang up and degrade performance. As production volume ramps, a fraction of the product always ends up nonsalable. In a robust design, the failing fraction may be so small that functional tests can be simplified, leading to further cost reductions. In contrast, designs sensitive to component tolerances require extensive testing and will suffer heavy yield losses. Reworking defective units incurs extra labor and parts charges, ultimately eroding profits.

Thus, redesigning to improve robustness in the face of normal manufacturing tolerances is a major challenge of moving from the engineering bench to mass production. This process is called *design for manufacturing (DFM)*.

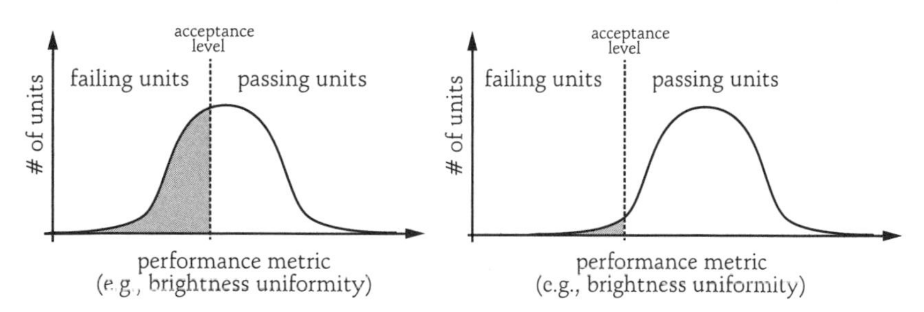

Left, before DFM, almost half the units are not meeting the acceptance level and are therefore failing. Right, after DFM, the acceptance level is the same, but the average performance is improved, leading to most units passing.

To understand the importance of DFM, consider these graphs. Each depicts a *bell curve*, which is an assumed statistical distribution of a particular parameter. The x-axis is a parameter of interest, and the y-axis is the number of items produced that hit the given parameter. For example, in a plot of the brightness of thousands of LEDs, the x-axis would be brightness, and the y-axis would be the number of LEDs that reach a given brightness. The position of the bell curve relative to the pass/fail criteria determines the net production yield.

On the right-hand curve, most LEDs are bright enough, and most of the production inventory is shippable. On the left-hand curve, maybe 40 percent of the LEDs pass. Given that most hardware companies operate with about a 30 to 50 percent gross margin, scrapping 40 percent of the material would mean the end of the business. In such a situation, the only viable options are to spend the time and effort to rework the LEDs until they pass or to lower the performance requirement. The product wouldn't be as high quality as hoped, but at least the business could keep operating.

Tolerances to Consider

The goal of DFM is to ensure that your product always passes muster and that you're never faced with the unsavory choice of reducing margins, lowering quality standards, or going out of business. But there are some component aspects to think about when applying DFM.

ELECTRONIC TOLERANCES

Passive component tolerances are the most obvious tolerances to design for. If a resistor's true value can be +/–5 percent of its labeled value, be sure the rest of your circuit can cope with the edge cases.

Active component datasheet parameters—like current gain (hFE) for bipolar transistors, threshold voltage (V_t) for field effect transistors (FETs), and forward bias voltage (V_f) for LEDs—can also vary widely. Always read the datasheet, and watch for parameters with a great disparity between their minimum and maximum values, a difference often referred to as a *min-max spread.* For example, the min-max on hFE for Fairchild's 2N3904 ranges from 40 to 300, and the V_f on a superbright LED from Kingbright is between 2 and 2.5V.

Nominal operating voltage aside, a component's maximum voltage rating is particularly important for capacitors and input networks. I try to use capacitors rated for twice the nominal voltage; for example, where possible, I use 10V capacitors for 5V rails and 6.3V capacitors for 3.3V rails. To understand why, consider ceramic capacitor dielectrics, which have reduced capacitance with increasing voltage. In designs operating near a ceramic capacitor's maximum voltage, that component's operating capacitance will be at the negative end of its tolerance range. Also, *input networks* (any part of the circuit that a user can plug something into) are subject to punishing electrostatic discharge and other transient abuses, so pay special attention to the ratings of capacitors there to achieve your desired reliability.

Finally, after you have a good sense of the components you'll use, pay close attention to trace widths and layer stack variations when designing your PCB. These will impact systems that require matched impedance or deal with high currents.

MECHANICAL TOLERANCES

Electronic tolerances aren't the end of your worries, though; mechanical tolerances are important, too. Neither PCBs nor cases will come out exactly the right size, so design your case with some wiggle room. If your case design has zero tolerance for the PCB dimensions, half the time the factory will force PCBs into cases, when either the PCB is cut a little large or the case comes out a little small. This can cause unintentional mechanical damage to the circuitry or the case.

And don't forget about cosmetic blemishes! Any manufactured product is subject to small blemishes, such as dust trapped in plastics, small scratches, sink marks, and abrasions. It's important to work out the acceptance criteria for such defects with the factory ahead of time. For example, you might tell the factory that a unit can be considered "good" if it has no more than two dot blemishes larger than 0.2mm, no scratch longer than 0.3mm, and so on. Most factories will have a particular system they've adopted to describe and enforce these standards. If you discuss these parameters in advance, the factory can craft the manufacturing process to avoid such defects, as opposed to the more expensive alternative of building extra units and throwing away those that don't meet criteria imposed late in the game.

Of course, avoiding defects isn't free. To keep your product cheaper, avoid high-gloss finishes and consider using matte or textured finishes that naturally hide blemishes.

Following DFM Helps Your Bottom Line

To imagine DFM in a real-world scenario, return to the bicycle safety flasher case study from "How to Make a Bill of Materials" on page 74. Say the prototype design calls for an array of three LEDs in parallel, each with its own resistor to set the current. The *forward bias voltage*, or V_f, of an LED at a given brightness can vary by perhaps 20 percent between devices; in this case, that swing is from 2.0 to 2.5V.

A design that limits the current to the LEDs with resistors, called *resistive current limiting*, will amplify this variation. This happens because an efficient circuit would drop a minority of the voltage across the current-limiting resistor, leaving the parameter that sets the current (the voltage drop across the resistor) more sensitive to the variation in V_f. Since the brightness of an LED is not proportional to the voltage but rather the current flowing through it, setting the LED brightness with resistive current limiting can cause jarring inconsistencies in LED brightness.

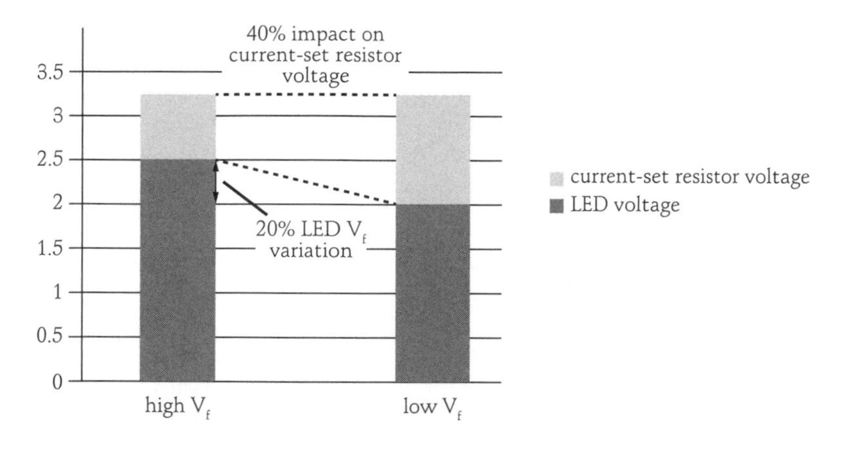

Comparing high V_f and low V_f corners

In this example, a 20 percent LED V_f variation (from 2.0V to 2.5V, per the LED manufacturer's specification) leads to a 40 percent change in the voltage across a current-set resistor for a fixed 3.3V supply. This will cause a 40 percent change in the current flowing through the LED. As brightness is directly proportional to current, the change manifests as up to a 40 percent variation in perceived brightness between individual LEDs. A design like that may work well most of the time; the problem would only be pronounced when a high V_f unit is observed next to a low V_f unit.

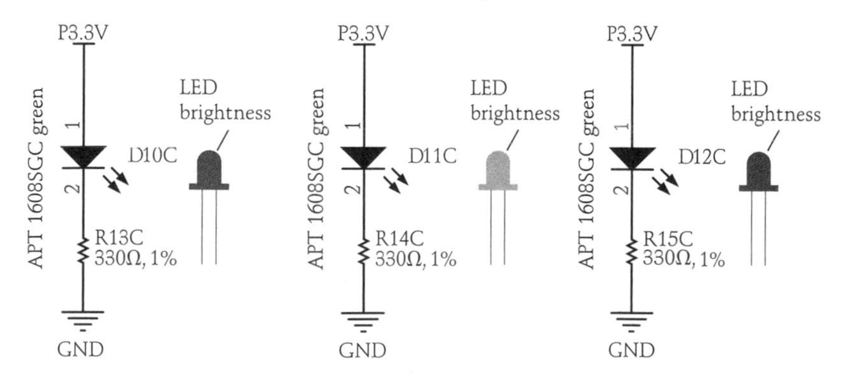

Setting current for individual LEDs using resistors can lead to dramatic variations in brightness.

The one or two units prepared on the lab bench during development may have looked great, but in production a meaningful fraction may have such serious brightness uniformity issues that units must be rejected. As most large hardware businesses have to survive on lean margins, losing even 10 percent of finished goods to defects is a terrible outcome.

One stop-gap option is to rework the failed units. A factory can identify an LED that is too dim or too bright in an array and replace it with one that better matches its cohorts. But that rework would drive up costs and result in an unexpected and unpleasant invoice at the 11th hour of a manufacturing program. Naive designers may be inclined to blame the factory for poor quality and argue over who should bear the cost, but it's better to proactively avoid these kinds of problems by subjecting every design to a DFM check and using a small pilot run to sanity-check yield before punching out a whole bunch of units.

The cost of yield fallout quantifies how much money to spend on extra circuitry to compensate for normal component variability. For example, a product with a $10 *cost of goods sold (COGS)* that yields 80 percent good units has an effective cost per salable unit of $12.50, as calculated with this formula:

$$\text{Effective cost} = \text{COGS} \times \text{total units built} / \text{yielded units}$$

Increasing the COGS by $2.50 to improve yield to 100 percent would allow you to break even. But using the same formula, spending $1 extra dollar in COGS to improve yield to 99 percent would actually improve the bottom line by $1.38.

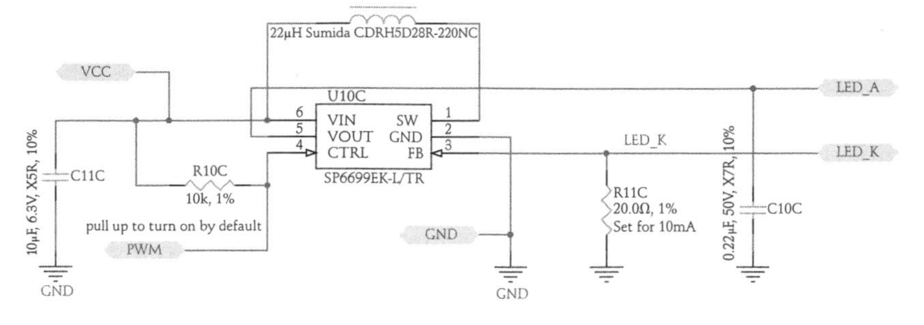

A circuit to set the current on three LEDs, created by applying DFM

In the case of the bicycle safety light, that dollar could be spent on a current-feedback boost regulator IC like the SP6699EK-L/TR, allowing the LEDs to be stacked in series instead of parallel. The design would be far more complicated and expensive than using individual resistors, but it would guarantee each LED has a consistent, identical current flowing through it by driving all three LEDs in a series circuit with a fixed-current feedback loop. That would virtually eliminate brightness variation. While the cost of the boost regulator is much greater than the penny spent on three current-limiting LEDs, the improvement in manufacturing yield more than pays for the extra component costs. In fact, this trick is standard practice for applications that require good uniformity of brightness out of LEDs, such as in the backlights of LCD panels. A typical mobile phone backlight uses about a dozen LEDs, but, thanks to circuits like this, you never see light or dark splotches despite the large variations in V_f between the constituent LEDs.

The Product Behind Your Product

Alongside dealing with tolerances, another often-neglected design responsibility is the test program. A factory can only detect the problems it is instructed to look for. Therefore, every feature of a product must be tested, no matter how trivial. For example, on a chumby device, every user-facing feature had an explicit factory test, including the LCD, touchscreen, audio, microphone, all the expansion ports (USB, audio), battery, buttons, knobs, and so on. I made sure that even the simplest buttons were tested. While it's tempting to skip testing such simple components, I guarantee that anything not tested will lead to returns.

I like to call the factory tester "the product behind your product." That's because in some cases, the factory tester is more complicated and more difficult to engineer than the product you're trying to sell. This is particularly true of simple products.

A REAL-WORLD TEST PROGRAM

As a case study, consider this microcontroller sticker from Chibitronics, a project I discuss at length in Chapter 8.

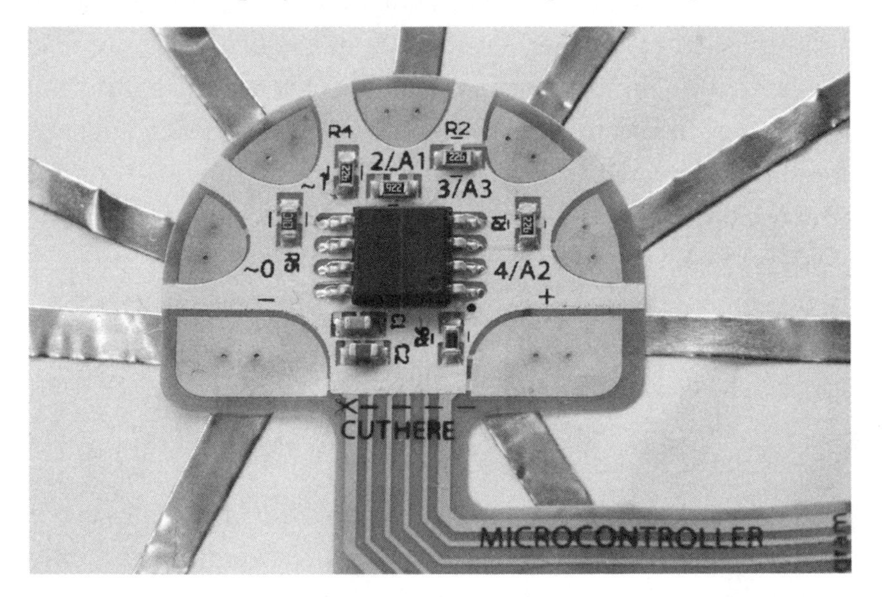

A microcontroller circuit—on a sticker

This circuit is very simple: it consists of just an 8-bit AVR microcontroller and a handful of resistors and capacitors.

(It's also the same product referred to in the ECO example on page 84.) My collaborator and I sketched in Adobe Illustrator for about two days before we derived the final shape for this product. Then we spent about a day in Altium designing the circuit, and about a week coding in the Arduino IDE to create its firmware. In all, the development process took about two weeks. For production, the microcontroller is paired with a set of sensors that can process sound, light, and touch, and as a result, the test program runs on all four at the same time.

The testing machine for the Chibitronics microcontroller sticker

The test rig pictured consists of a 32-bit ARM computer running Linux with a graphical UI rendered on an HDMI monitor. Behind this is an FPGA, some adapter electronics to create analog waveforms for testing, and a mechanical pogo-pin assembly for touching down on the sticker. Breaking down the design process for this rig into its component parts, we spent:

- Several days designing in Altium
- A week programming in the Xilinx ISE for the FPGA
- A couple of weeks hacking on Linux drivers
- A couple of solid months hacking in C++, to create the Qt integration framework
- A couple of days in SolidWorks, to create the mechanical apparatus to hold the whole thing together

Altogether, creating the tester for the microcontroller sticker took over two months, compared to the two weeks to create the product itself.

Why go through all this effort? Because time is money, and defects and returns are expensive to process. The tester can process one board in under 30 seconds; and in those 30 seconds, the tester has to program two microcontrollers; test sensors for light, sound, and touch; and confirm operation at both 5V and 3V. A manual test for all these operations could take several minutes of skilled labor and wouldn't be as reliable. Thanks to this tester, we processed zero returns due to defective material. Also, the graphical UI on the tester makes it very easy for the factory to determine exactly which point in the circuit is failing, facilitating fast rework of any imperfect material.

GUIDELINES FOR CREATING A TEST PROGRAM

As a rule of thumb, for every product you make, you're actually making two related products: one for the end user, and a test for the factory. In many ways, the test for the factory has to be as user-friendly and foolproof as the product itself; after all, tests are not run by electrical engineers. But the related testing product will be much quicker and faster to build if adequate testing features are designed into the consumer product.

And no, don't outsource the test program to the factory, even if the factory offers that service. The factory often won't understand your design intent, so their test programs will either be inefficient or test for the wrong behavior. Factories also have an incentive to pass as much material as possible, as quickly as possible, so their test programs tend to be primitive and inadequate.

Here are some guidelines to follow when designing your own program:

Strive for 100 percent feature coverage.

Don't overlook simple or secondary features like status LEDs or an internal voltage sensor. When creating the test list, I take an "outside/inside" approach. First, look at the product from the outside: list every way a consumer can interact with it. Does your test program address every interaction surface, even if only superficially? Is every LED lit, every button pressed, every sensor stimulated, and every memory device touched? Has every bullet point in your marketing material been confirmed? Promising "world-class" RF sensitivity is different from simply advertising the presence of a radio. Then, think about the inside: from the schematic, look at every port and consider key internal nodes to monitor. If the product has a microcontroller, review which drivers are loaded to cross-check the test list, and make sure no components are forgotten.

Minimize incremental setup effort.

Optimize the amount of time required to set up the test for each unit. This is often done through jigs that employ pogo pins or prealigned connector arrays. A test that requires an operator to manually probe a dozen test points with a multimeter or insert a dozen connectors is time-consuming and error-prone. Most factories in China can help design the jig for a nominal cost, but jig design is easier and more effective if the design itself already includes adequate test points.

Automate test procedure into a linear flow.

An ideal test runs with a single button press, and produces a pass or fail result. In practice, there are always stop points that require operator intervention, but try not to require too much. For example, don't require an operator to key in or select an SSID from a list during each Wi-Fi connectivity test. Instead, fix the test target's SSID and hardcode that value into a test script so the connection cycle is automatic.

Use icons and colors, not text, to communicate with operators.

Not every operator is guaranteed to be literate in a given language.

Employ audit logs.

Record test results correlated to device serial numbers by incorporating a barcode scanner into the test rig. Alternatively, have the device print a coupon with a unique, timestamped code or a locally stored audit log to prove which units passed a test. Logs will help you figure out what went wrong when a consumer returns a failed product, and they let you quickly check that all products were tested. After an eight-hour shift of testing, an operator may make mistakes, such as accidentally putting a defective unit into the "good" bin. Being able to check that every shipped product was subjected to and passed the full test can help you identify and isolate such problems.

Provide an easy update mechanism.

Like any program, test programs have bugs. Tests also need to evolve as your product is patched and upgraded. Have a mechanism to update and fix test programs without visiting the factory in person. Many of my test fixtures can "phone home" via a VPN, and I can SSH into the jig itself to fix bugs. Even my simplest jig employs a Linux laptop (or equivalent) at its core. This is in part because Linux is easier to update and maintain than a bespoke microcontroller that requires a special adapter for firmware updates.

These guidelines are easy to implement if your product is designed with testability in mind. Most of the products I design run Linux, and I leverage the processor inside the product itself to run most tests and help manage the test user interface. For products that lack user interaction surfaces, an Android phone or a laptop connected via Wi-Fi or serial can be used to render the test user interface.

Testing vs. Validation

Production tests are meant to check for assembly errors, not parametric variations or design issues. If a test is screening out devices because of normal parametric component variations, either buy better components or redo your design.

For consumer-grade products, you don't need to run a five-minute comprehensive RAM test on every unit. In theory, your product should be designed well enough that if it's all soldered together correctly, the RAM will do its job. A quick test to check that there are no stuck or open address pins is often enough. Name-brand chip vendors typically have very low defectivity, so you're not validating the silicon; rather, you're validating the solder joints and connectors and checking for missing or swapped components. (But if you buy clone chips or off-brand, remarked, or partially tested devices to cut costs, I recommend making a mini validation program for those components.)

VALIDATING A SWITCH

To illustrate the difference between production testing and validation, let's look at how both might work for a switch.

A production test for a switch may simply ask the operator to hit the switch a few times and verify that the feel is right, and that electrical contact is made through a simple digital indicator. A validation test, on the other hand, may involve selecting a few devices at random, measuring the switch contact resistance with a multimeter that is accurate to five significant digits (also called a *five-digit multimeter*), subjecting the devices to elevated humidity and temperature for a couple of days, and then putting the devices into an automated jig that cycles the switches 10,000 times. Finally, you might remeasure the switch contact resistance with a five-digit multimeter and note any degradation in close-state contact resistance.

Clearly, this level of validation can't be performed on every device manufactured. Rather, the validation program evaluates the switch's performance over the expected lifetime of the product. The production test, on the other hand, just makes sure the switch is put together right.

NOTE *It's good practice to rerun validation tests on a couple of randomly sampled units out of every several thousand units produced. There are formulas and tables you can use to compute how much sampling you need to achieve a certain level of quality; just search online for "manufacturing validation test table."*

But how much testing is enough? You can derive one threshold for testing through a cost argument. Every additional test run incurs equipment costs, engineering costs, and the variable cost of the test time. As a result, testing is subject to diminishing returns: at some point, it's cheaper just to take a product return than to test more. Naturally, the testing bar is much higher for medical or industrial-grade equipment, as the liability associated with faulty equipment is also much higher. Likewise, a novelty product meant to be given away may need much less testing.

DESIGNING YOUR TEST JIG

A final thought: always apply solid engineering to your test jig design. When I worked on the chumby 8, there was a problem where a 50-pin flat flex cable adapter was exhibiting random cold-solder-joint failures. I asked the factory to build a test to validate the adapters. Their solution was to hang LEDs from every pin of the adapter, apply a test voltage to one side of the cable, and look for LEDs that didn't light on the other side. The cold solder joints weren't simply open or closed; some just had high resistance. Enough current would flow to light an LED, yet there was also enough resistance to cause a fault in the design.

The factory proposed buying 50 multimeters and attaching them to every pin to check the resistance manually, which would have been expensive and error-prone. It's not reasonable to expect an operator to look at 50 displays hundreds of times a day and be able to reliably find the out-of-spec numbers. Instead, I chose to daisy-chain the connections across the adapter and use a single multimeter to check the net resistance of the daisy chain. By putting the connections in series, I could check all 50 connections with a single numeric measurement, as opposed to the subjective observation of an LED's brightness.

As this case illustrates, there are good and bad ways to implement even a test as simple as checking for cold solder joints on a cable adapter. Ever more complicated components require ever more subtle tests, and there's real value in using engineering skills to craft efficient yet foolproof tests.

FINDING BALANCE IN INDUSTRIAL DESIGN

Even if your product passes all validation tests with flying colors, it still may not be successful if consumers don't want it. Remember: sex sells. To within a factor of two or so, the performance of a CPU or amount of RAM in a box is less important to a typical consumer than how the device looks. Apple devices command a hefty premium in part because of their slick industrial design, and many product designers aim to emulate the success of Sir Jonathan Ive, Apple's chief design officer, in their own products.

There are many schools of thought in *industrial design*, the process of designing how a product will look before actually making it. One school invokes the monastic designer, who creates a beautiful, pure concept, and the production engineers, who spoil the design's purity when they tweak it for functionality. Another school invokes the pragmatic designer, who works closely with production engineers to hammer out gritty compromises to produce an inexpensive and high-yielding design.

In my experience, neither extreme is compelling. The monastic approach often results in an unmanufacturable product that is either late to market or expensive to produce. The pragmatist approach often results in a product that looks and feels so cheap that consumers have trouble assigning it a significant value. The real trick is understanding how to strike a balance between the two, and it begins by getting into the factory and understanding how things are done. Here's a couple of examples of what I've learned about how different factory processes affect that balance, from Chumby and Arduino.

The chumby One's Trim and Finish

Trim and finish are difficult, making them points of distinction in a product's appearance. When I worked at Chumby, we wanted the final product to have a minimalist, honest finish. (*Honest finishes* feature the natural properties of the material systems in play and eschew the use of paints and decals.) Minimalist designs are very hard to manufacture because with fewer features, even tiny blemishes stand out. Honest finishes can be difficult, too, as all the burs, gates, sinks, knits, scoring, and flow lines that are facts of life in manufacturing are laid naked before the consumer. As a result, this school of design requires well-made manufacturing tools that are constantly checked and maintained throughout production.

If you don't have pockets deep enough to invest in new equipment and capabilities on behalf of your factory (that is, if you're not a *Fortune* 500 company), the first step is to learn the vocabulary available. A *design vocabulary* is defined by the capabilities of the factory or factories producing the goods, like what materials you can obtain, what finish is possible, what tolerances are achievable, and what fastening technology exists. These are all heavily dependent upon the processes available to your factory.

Therefore, I find that visiting a factory in person early in the design process results in a better design. After a factory visit, you'll discard some design vocabulary, but you'll discover some new vocabulary as well. The engineers who work in the factory day in and day out develop process innovations that can open up novel design possibilities that you won't discover unless you visit.

The chumby One is a concrete example of the impact manufacturing processes can have on design outcome. In the original concept art, a blue highlight was added around the front edge to resemble a speech balloon, like those used in comic strips. The idea was that the chumby would caption your world with snippets from the internet.

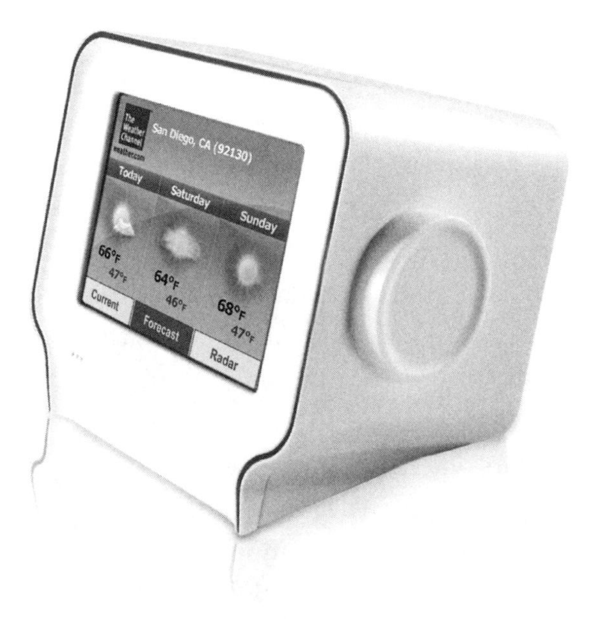

A finished chumby One unit

But applying a blue trim across a raised surface was very hard. The first factory used paint, because the front edge wasn't flat enough to make silk screening an option. *Pad printing* (also known as *tampo printing*, a process in which ink is transferred from a silicone pad to an object) can handle curved surfaces, but the alignment of the ridge on the chumby One wasn't good enough, and the tiniest ink bleed over the edge looked terrible from the side. Decals and stickers likewise couldn't achieve the alignment we wanted. In the end, a small channel was carved to contain the paint, and the factory created the highlight with a stencil and spray paint.

The yield was terrible. In some lots, over 40 percent of the chumby One cases were thrown away due to painting errors. Fortunately, plastic is cheap, so throwing away every other case after painting had a net cost impact of about $0.35.

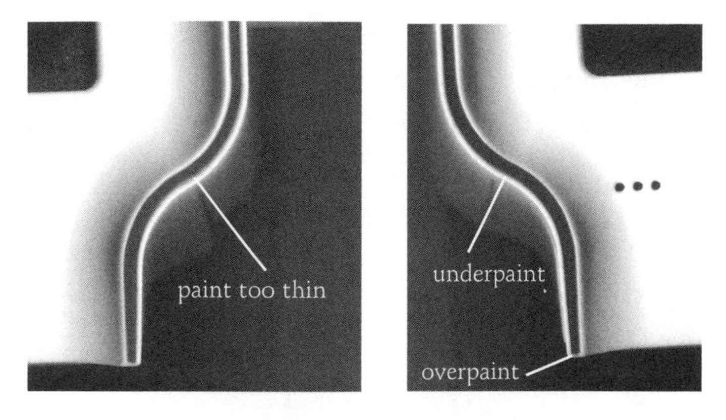

Two chumby One units with bad paint jobs

Midway through production, we started producing chumby One units in a second-source facility. The second factory had different plastic molding equipment, and unlike the first factory, this facility could do *double-shot molds*. A double-shot mold involves twice the number of tools of a single-shot injection mold, but it can injection-mold two different colors, or even two different materials, into the same mold. At the new factory, we tried a double-shot process instead of painting for the thin blue strip.

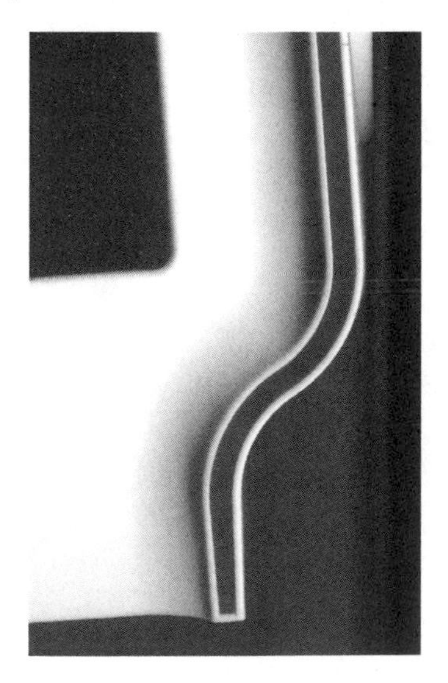

A perfect chumby One ridge,
from the double-injection mold process

The results were stunning. Every unit came off the line with a crisp blue line, and no paint meant a cleaner, more honest finish. But the cost per case jumped to $0.94 apiece with the more expensive process, despite the 100 percent yield. It would have been cheaper to throw away more than half of the painted cases, but even the best painted cases could not compare to the quality of the finish delivered by the double-shot tool.

The Arduino Uno's Silkscreen Art

Another great example of how tweaking a factory process can improve a product's appearance is the Arduino motherboard. The wonderfully detailed artwork on the back side, sporting an outline of Italy and very fine lettering, isn't silkscreen. The factory that makes these boards actually puts on two layers of soldermask: one blue and one white.

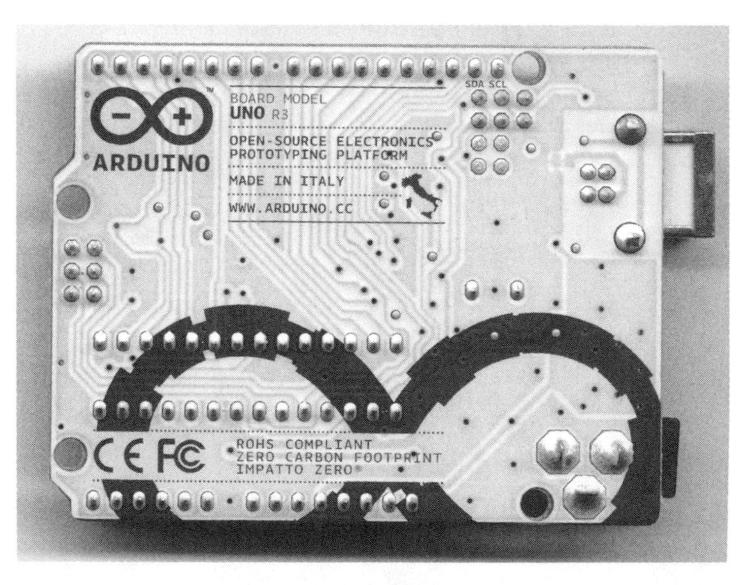

The underside of an Arduino Uno R3

When Arduino boards are manufactured, soldermask is applied through the photolithographic process I described in "Where Arduinos Are Born" on page 44. This process results in artwork with much better resolution, consistency, and alignment than a silkscreen. And since an Arduino's look is the circuit board, this art gives the product a distinctive, high-quality appearance that is difficult to copy using conventional processing methods.

Thus, the process capability of a factory (whether it's painting versus double-shot molding, or double soldermasking versus silkscreening) can have a real effect on a product's perceived quality, without a huge impact on cost. The factory, however, may not appreciate the full potential of its processes, and until a designer interacts with the facility directly, your product can't harness that potential, either.

Unfortunately, many designers don't visit a factory until something has gone wrong. At that point, the tools are cut, and even if you discover a cool process that could solve all your problems, it's often too late.

My Design Process

Design is an intensely personal activity, and as a result, every designer will develop their own process. If you need a framework for developing your own, however, this is the general process I might use to develop a product on a tight, startup budget:

1. Start with a sketchbook. Decide on the soul and identity of the design, and pick a material system and vocabulary that suits your concept. But don't fall in love with it, because it may have to change.
2. Break down the design by material system, and identify a factory capable of producing each material system.
3. Visit the facility, and note what is actually running down the production lines. Don't assume anything based on the one-off units from the sample room. Practice makes perfect, and from the operators to the engineers, factory workers execute procedures they do daily much better than they would an arcane capability they don't use often.

4. Reevaluate your design based on a new understanding of what's possible at the factory, and iterate. Go back to step 1 if small tweaks aren't enough. This is the stage when it's easiest to make compromises without sacrificing the purity of your design.
5. Rough out the details of your design. Pick sliding surfaces, parting lines where pieces of the case snap together, finishes, fastening systems, and so on based on what the factory can do best.
6. Pass a revised drawing to the factory, and work with them to finalize details such as draft angles, fastening surfaces, internal ribbing, and so on.
7. Validate the design using a 3D print and extensive 3D model checks.
8. Identify features prone to tolerance errors, and trim the initial manufacturing tool so that the tolerance favors modifications that will help you minimize costly changes to the tool. For example, consider injection molding, where a steel tool is the negative of the plastic it's molding. Removing steel from a tool (adding plastic) is easier than adding steel (removing plastic), so target the initial test shot to use more steel on critical dimensions, as opposed to too little. A button is one mechanism that benefits from tuning like this: predicting exactly how a button will feel from CAD or 3D prints is hard, and perfecting the tactile feel usually requires a little trimming of the tool.

Of course, this process isn't a set of hard rules to follow. You may need to add or repeat steps based on your experience with your factory, but if you choose a good factory, this should be a good starting point.

PICKING (AND MAINTAINING) A PARTNER

Just like the wands from *Harry Potter*, a good factory chooses you as much as you choose it, so forget the term *vendor* and replace it with *partner*. If you're doing it right, you aren't simply instructing the factory; there should be a frank dialogue about the trade-offs involved and how the manufacturing process can be improved. That's the only way to get the best product possible.

A healthy relationship with a factory can also lead to better payment terms, which improves your cash flow. In some cases, factory credit can directly replace raising venture capital, taking loans, or getting Kickstarter funding. As a result, I treat good factories with the same respect as investors and partners in a business. For an idea of what that means, here are some tips on how to choose and work with your factory.

Tips for Forming a Relationship with a Factory

First, pick the right-sized factory for your product. If you work with a factory that's too big, you risk getting lost in bureaucracy and pushed out of the production line by bigger customers at critical times. Work with a factory too small, and it won't be able to provide the services you need. As a rule, I pick the biggest facility where I can get direct access to the *lao ban* (factory boss) on a regular basis, because if you can't talk to the boss, you're nobody. It's a good sign if the lao ban is there on the first meeting to give you a tour and asks astute questions about your business over lunch.

Second, follow the adage "Sunlight is the best disinfectant." If a factory won't quote with an open BOM, where the cost of every component, process, and margin is explicitly disclosed, I won't work with them. Cost reduction discussions cannot function without transparency, because there are too many places to bury costs otherwise. Likewise, if cost discussions turn into a game of whack-a-mole, where reduced costs on one line item are inexplicably popping up in another, run away.

This final tip applies primarily to startups. In your early stages, everyone knows your cash supplies are finite. Even if you've just closed a big round of financing, swaggering into a factory with money bags is not a sustainable approach. Smart factories know your cash supplies are limited, and if the greatest value you propose to bring to the factory is piles of money, your value is limited; in the best case, it won't really pay out until years down the road when the product is shipping in high volumes. As a result, it's helpful to try to deliver value to the factory in nonmonetary ways.

As silly as it sounds, being a pleasant and constructive person goes a long way in currying the favor of your facility. Manufacturing is a high-stress, low-margin business, and everyone in the facility has to deal with difficult problems all day. I find I get better service—even better than customers with deeper pockets—if I treat my factories as I would treat a friendly acquaintance, and not as slave labor or a mere subcontractor. Mistakes happen, and being able to turn a bad situation into a learning experience will benefit you on the day you make a stupid (and perhaps expensive) mistake.

Tips on Quotations

Openness aside, know that if a quote seems too good to be true, it often is. When negotiating prices with a factory, step back and check if the quote makes sense. Factories that lose money on a deal will stop at nothing to make it back, and many manufacturing horror stories have roots in unhealthy cost structures. A factory's first prerogative is survival, even if that means mixing defective units into lots to boost margin, or assigning novice engineers to a flagging project to better monetize their seasoned engineers on more profitable customers.

As you evaluate a quote, make sure it includes the following:

- The price of each part
- The excess material for the job due to *minimum order quantities (MOQs)*
- Labor costs
- The factory's overhead cost
- *Nonrecurring engineering (NRE)* fees

Let's look at a few of these items in detail.

KEEPING AN EYE ON EXCESS

Excess is the result of what I call the "hot dogs and buns" problem. Hot dogs come in packs of 10, but buns come in packs of 8. Unless you buy 40 servings, you'll have leftover buns or hot dogs.

Likewise, many components only come in 3,000-piece reels. A 10,000-piece build requires 4 reels for a total of 12,000 pieces, leaving 2,000 pieces of excess. Factories can buy parts in cut tape or partial reels, but the cost per part of cut tape is much higher, as the risk of excess material is shifted onto the distributor.

Excess isn't all bad, though: it can be folded into future runs of a product. As long as your product sustains a decent production rate, excess component inventory should turn into cash on a regular basis. At some point, however, production will end or pause, and the bill for the excess will arrive, putting a crimp on cash flow. If a quote lacks an excess column, the factory may charge you for the full reel but keep the excess for their own purposes; this is where many of the gray-market goods in Shenzhen come from. They may also just send an unexpected invoice for it down the road, which often arrives at the worst possible time—revenue from the product has already ceased, but bills keep coming in. Either way, it's best to know up front the complete cradle-to-grave business model.

FIGURING OUT LABOR COSTS

Labor costs are devilishly tricky to estimate, but the good news is that for high-tech assemblies, labor is typically a small fraction of total cost. The labor cost of assembling small volumes of a straightforward board with 200 parts may be about $2 or $3 in China, while the cost of assembling in the United States is closer to $20 or $30. Even if labor prices double overnight in China and halve in the United States, China may still be competitive.

This is in contrast to the lower-value goods moving out of China (such as textiles), where the base value of the raw material is already low, so labor costs are a significant portion of the final product cost. I usually don't argue much over labor costs, since the end result of scrimping on labor is often lowered quality, and pushing too hard on labor costs can force the factory to reduce the workers' quality of life by trimming benefits.

THE FACTORY'S OVERHEAD

Negotiating factory margin is also a bit of an art, and there are no hard-and-fast rules. I'll give guidance here, but there are always exceptions to the rule, and every factory can cut you a special deal depending on the circumstances. Ultimately, it's important to look at the big picture when reviewing a factory's quote and use some common sense.

What constitutes a fair margin for a factory depends on how much value it adds to your product, and the volume of production. The definition of "margin" also varies depending on the facility. Some facilities include scrap, handling overhead, and even research and development expenses in the margin, while others may break those out on separate lines.

In general, margin ranges between single-digit and low double-digit percentages, depending upon volume, value add, and project complexity. For very low-quantity production lots (fewer than 1,000 pieces), you may also be charged a per-lot *line fee*. This fee partially defrays the cost of setting up an assembly line only to tear it down after a couple of hours. A line's throughput may be very fast, producing hundreds to thousands of units a day, but it also takes days to set up.

NONRECURRING ENGINEERING COSTS

NRE costs are onetime fees required to set up a production run, such a stencils, SMT programming, jigs, and test equipment. Note that reusing test equipment between customers is considered bad practice; if a multimeter is required as part of a production test, don't be surprised if a bill for a multimeter is tacked onto the NRE. Customers have drastically varying standards around the maintenance and use of test equipment, so good factories don't take chances with it.

Miscellaneous Advice

Who you can talk to and how open the factory is about costs are certainly key concerns, but with experience, you'll learn a lot more about dealing with factories that doesn't fall into any particular category. To close, here are a few more important points to keep in mind when selecting a factory.

SCRAP AND YIELD

Ideally, you'd pay a factory only for good, delivered items, and the factory would bear the burden of defective units. This gives the factory an incentive to maintain a high production quality, because every percent of defectiveness eats away at its margin. But if your design has a flaw or is too hard to build, and defectiveness is high, the factory may start shipping lower-quality units as a desperate measure to meet production and margin targets. It may also start selling defective goods on the gray market to recover cost, leading to brand reputation problems down the road.

To avoid situations like that, reach an understanding with the factory ahead of time on how to handle scrap units or exceptional yield loss. This may include, for example, a dedicated "scrap" line item inside the quotation to handle defectiveness explicitly.

ORDER MORE UNITS THAN THE PROVEN DEMAND

Despite everyone's best efforts, mistakes will happen, customers will receive bad devices, and you'll want extra working units for returns and exchanges. Ordering 1,000 pieces to fulfill a 1,000-piece Kickstarter campaign means if customers want to return or exchange units that were broken in shipping, all you can do is issue refunds. It's just not practical to fire up the factory to make a dozen replacement units.

As a general rule, I order a few percent excess beyond the number of units I need to deliver to customers, to have stock on hand to handle returns and exchanges. Units that don't get used up by the returns process can be turned into demo loaners or business development giveaways to drum up the next set of orders!

SHIPPING COSTS MONEY

Keep an eye on shipping costs. These fees aren't typically built into a factory's quotation, but they impact your bottom line, even more so for low-volume products. Shipping FedEx is a great way to save time, but it's also very expensive. Courier fees can easily wash out the profit on a small project, so manage those costs.

NOTE *Couriers offer discounts to frequent shippers, but you have to call in to negotiate the special rates.*

FACTOR IN IMPORT DUTIES

Components imported to China without an import license are levied a roughly 20 percent compulsory duty on their value. The general rule for China is dutiable on import, duty free on export. If something is accidentally shipped across the border to Hong Kong, expect to pay a duty to get it back into China, too.

Get a customs broker to work angles for saving money; for example, some brokers can get goods taxed by their weight and not their value, which for microelectronics is typically a good deal. I haven't figured out all the customs rules, as they seem to be a moving target. Every month it seems there's a new rule, fine, exceptional fee, or tariff to deal with. There are also plenty of shady ways to get goods into China, but I sleep better at night knowing I do my best to comply with every rule.

Quotations don't include duties, because factories assume by default that you will have an import license. Import licenses enable the duty-free import of goods. But import licenses cost a few thousand bucks, take weeks to process, and have no room for flexibility, as they are tied to an exact BOM for the product. Small engineering change orders can invalidate an import license. I've known customs officers to count the number of decoupling caps on a PCB, and if it doesn't match the count in the license, a fine is levied and the license is invalidated. Even deviations in the material used to line a decorative box can invalidate a license. In short, this import license scheme favors high-volume products, and punishes low-volume producers, so tread lightly.

CLOSING THOUGHTS

Going to China for manufacturing clearly isn't for everyone. Particularly if you're based in the United States, the overhead of courier fees, travel, duties, and late-night conference calls adds up rapidly. As a rule of thumb, a small US-based company is often better off assembling PCBs in the United States for volumes under 1,000 units, and you won't start seeing clear advantages until volumes of perhaps 5,000 to 10,000 units.

That math shifts in China's favor as processes like injection molding and chassis assembly come into play, due to the expertise Chinese factories have in these labor-intensive processes. The break-even point can also be much lower if you live in or near China, as courier fees, travel, and time-zone impact are all a small fraction of what they'd be from the United States. This compounds with the fact that locals are more effective at leveraging the component ecosystem in China, leading to further cost reductions compared to a design produced using only US parts.

On the other hand, physically large assemblies or systems built using lots of dutiable components may be cheaper to build domestically, as they save on shipping costs and tariffs. In the end, keep an open mind and try to consider all the possible secondary costs and benefits of domestic versus foreign manufacturing before deciding where to park production.

Part 2

thinking differently: intellectual property in china

China has a reputation for lax enforcement of intellectual property (IP) laws, and that leads to problems like fake and copycat products. This part of the book takes a nuanced look at China's IP ecosystem and finds a novel way to reward innovation that serves as an alternative to traditional Western IP practices.

First, consider this question: what, exactly, constitutes a fake? It seems relatively straightforward to answer; anything that's not an original must be a fake. The situation becomes muddied, however, when you consider the possibility that some contract manufacturers produce fakes by running a *ghost shift*, an after-hours production run not reported to the product's brand owner. These items are produced on the same equipment, by the same people, and with the same procedures as the original product, but they're sold directly to customers at a much higher margin to the manufacturer.

In fact, the spectrum of fakes runs an entire gamut of possibilities. Used and damaged goods get upcycled; production rejects with minor flaws are refurbished and sold as originals; original products get relabeled to advertise a higher capability or capacity (for example, memory cards with 4GB actual capacity are sold as 8GB), and so on. Chapter 4 relates several encounters I've had with fake goods in China, and dives into the issues and incentives enabling the rise of such fakes.

Cloning and copying are also common practices in China. A nebulous and sometimes shadowy group of rogue innovators known as *shanzhai* creates products that attempt to mimic the features and function of an original product, often with assistance from the original's blueprints. But the clones are heavily modified to save cost or include unique features. Often, the most offensive aspect of the practice is the use of the original product's brands and trade dress on the clones. Aside from trademark violations, a look inside the products reveals an incredible amount of original engineering and innovation.

Dismissing the shanzhai as mere thieves and copycats overlooks the fact that they can achieve what few Western companies can: they can build complete mobile phones, and on a shoestring budget to boot. Chapter 5 takes a deep dive into a prime example of shanzhai engineering, a feature phone designed for emerging markets that costs under $10. The phone is a tour de force of cost reduction and a fresh look at ways of building to address markets that are untouchable with Western engineering practices.

One of the most insightful lean engineering practices enabling the creation of complex systems on a shoestring budget is the shanzhai method for sharing IP. I'll explore this by comparing and contrasting the Western notion of open source with the shanzhai method, which I refer to as *gongkai*. In Western law, open source has a formal definition, referring specifically to an IP sharing system governed by an explicit license to share. This license is granted by the copyright holder, often with significant commercial restrictions. Open source advocates vigorously defend this notion and are quick to disavow any IP that doesn't explicitly use an approved license.

In gongkai, if you can obtain a copy of the blueprints, you can use them as you please; it doesn't matter who made them. Yet people still share their ideas because the blueprints act as an advertisement. Blueprints often refer explicitly to certain chips or contain contact information for the firm that drew them. The creators hope circulating their blueprints will bring business to their factory when people order parts or subassemblies referenced within, or when people call their firm to improve or customize the design. In other cases, blueprints are traded. For example, there are bulletin board exchanges where before you download a blueprint, you must contribute one of your own.

In short, the gongkai IP ecosystem is a variant of the ad-driven business model, but optimized for hardware-oriented businesses. Just as Google provides high-quality search, email, and mapping services for free in exchange for showing ads, shanzhai innovators share ideas to land follow-up orders in their factories.

Here lies a key distinction between most Western innovators and their counterparts in Shenzhen: everyone who is anyone in Shenzhen owns or has close ties to a factory. The fastest path to material wealth is selling more product. Arguing over who has rights to abstract ideas is a waste of effort best left for *baijiu*-fueled discussions after dinner.* On the opposite end of the spectrum are Western patent trolls so removed

* *Baijiu* is a type of strong Chinese alcohol.

from factories that they probably don't even have a soldering iron, yet they invest millions of dollars into litigation and collecting royalties on ideas they didn't invent.

Neither system is perfect, but the gongkai method is uniquely adapted to the fast pace of technology. In a world where chips get faster and cheaper every couple of years, a 20-year patent lifetime is an eternity. Spending a decade to bring a product to market simply is not an option; the best factories in China can turn a napkin sketch into a prototype in days and bring it to scale production in weeks. Long patent terms may be appropriate for markets like pharmaceuticals, but in fast-moving markets, investing months and tens of thousands of dollars in lawyer fees to negotiate a license or just apply for a patent can lead to missed opportunities.

Perhaps a discussion on reforming the Western patent system is long overdue. The gongkai ecosystem is living proof that granting 20-year monopolies on ideas as trivial as "slide to unlock" for a smartphone may not be the One True Path to incentivize innovation. I look forward to starting the conversation with this whirlwind tour of the good, the bad, and the ugly of the Chinese IP.

4. gongkai innovation

If the term *intellectual property* sounds like an oxymoron to you, you're not alone. If I give you an apple and say, "This is your apple," what that means is pretty clear. You can do what you want with that apple: you can eat it, sell it, or even use the seeds to plant an apple tree and make more apples, which you can then sell or use to feed your family. But if I hand you a phone and say, "This Apple iPhone is yours," you own the collection of atoms in your hand, but you have extremely limited rights to the software, patents, and trademarks—the intellectual property—associated with that phone. Unlike with the fruit, you can't take what's inside your iPhone and use that knowledge as a seed to make more iPhones.

Intellectual property works very differently in China, though. There, you could (and people do) use a phone as the seed for your own original works. Two experiences I had in China opened my eyes to the fact that there isn't one true path for dealing with intellectual property.

I BROKE MY PHONE'S SCREEN, AND IT WAS AWESOME

My first story begins, as many of my adventures do, with stepping out of a taxi at the Futian border checkpoint going into China. It was May 2014, and I was heading to Shenzhen to hammer out production plans for the Novena open hardware laptop, which I'll talk more about in Chapter 7. As I stepped out of the taxi, my hand caught on my backpack, sending my phone tumbling toward the concrete sidewalk. As the phone smashed into the ground, I heard the dry "thud" of a shattering touchscreen.

There is no better place in the world to break your phone's screen than the border crossing into Shenzhen. Within an hour, I had a new screen installed by skilled hands in Hua Qiang Bei, for just $25—including parts and labor.

I originally planned to replace the screen myself. The phone still worked, so I hastily visited iFixit for details on how to replace the screen and then booked it to Hua Qiang Bei to purchase replacement parts and tools. The stall I visited quoted me about $120 USD for a new screen, but then the shop owner grabbed my phone out of my hands and launched a built-in self-test program by punching *#0*# into the dialer UI.

She confirmed that there were no bad pixels on my OLED display and that the digitizer was still functional, just cracked. She then offered to buy my broken OLED and digitizer module, but only if her shop could replace my screen. I said that would be fine as long as I could watch to make sure they didn't swap out any other parts.

Of course, they had no problem with that. In 20 minutes, they took my phone apart, removed the broken module, stripped the adhesive from the phone body, replaced the adhesive, fitted the phone with a "new" (presumably refurbished) module, and put it all back together. The process involved a hair dryer (used as a heat gun), copious amounts of contact cleaner (used to soften the adhesive), and a very long thumbnail (in lieu of a spudger/guitar pick). Unfortunately, I couldn't take pictures of the process because the device I would have used to do so was in pieces in front of me.

This is the power of recycling and repair. Instead of paying $120 for a screen and throwing away a functional piece of electronics, I just paid the cost to replace the broken glass. I had assumed that the glass on the digitizer was inseparable from the OLED, but apparently those clever folks in Hua Qiang Bei found an efficient way to recycle those parts. After all, the bulk of the module's cost was in the OLED display. The touchscreen sensor electronics, which were also grafted onto the module, were undamaged by the fall. Why waste perfectly good parts?

And so my phone had a broken screen for all of an hour, and it was fixed for less than the cost of shipping spare parts to Singapore (my country of residence). Experiences like this get me thinking: why aren't there services like this in every country? What makes Shenzhen so unique that you can go from a broken screen to a fixed phone in half an hour for much less than the cost of a monthly phone bill? A multitude of factors contribute to this phenomenon, most of which can be traced to a group of people called the *shanzhai*.

SHANZHAI AS ENTREPRENEURS

The shanzhai of China originally became famous as the producers of knockoffs of products like the iPhone, so they've historically been dismissed by the popular press as simply "copycat barons." But I think they may have something in common with teams like Hewlett and Packard or Jobs and Wozniak, back when they were working out of garages.

Who Are the Shanzhai?

To understand why I think this, it helps to understand the cultural context of the word *shanzhai*. Shanzhai (山寨) comes from the Chinese words *mountain fortress*, but the literal translation is a bit misleading. The English term *fortress* connotes a large fortified structure or stronghold, perhaps conjuring imagery of castle turrets and moats. On the other hand, its denotation states that it is simply a fortified place, and this is closer to the original Chinese meaning, which refers to something like a cave or guerrilla-style hideout.

In its contemporary context, *shanzhai* is a historical allusion to the people who lived in such hideouts, like Song Jiang and his 108 bandits, a group of outlaws who lived in the 12th century. A friend of mine described Song Jiang as a sort of Robin Hood meets Che Guevara. He was a rebel and a soldier of fortune, yet selfless and kind to those in need. The tale is still popular today; my father instantly recognized it when I asked him about it.

Modern shanzhai innovators are rebellious, individualistic, underground, and self-empowered—just like Song Jiang. They're rebellious in the sense that they are celebrated for their copycat products. They're individualistic in the sense that they have a visceral dislike for the large companies. (Many shanzhai are former employees of large companies, both American and Asian, who departed because they were frustrated by the inefficiency of their employers.) They're underground in the sense that once a shanzhai "goes legit" and does business directly through traditional retail channels, they no longer belong to the fraternity of the shanzhai. They're self-empowered in the sense that they're universally tiny operations, bootstrapped on minimal capital, and their attitude is, "If you can do it, then I can as well."

An estimated 300 shanzhai organizations were operating in Shenzhen in 2009. Shanzhai shops range from just a couple of folks to a few hundred employees. Some specialize in processes like tooling, PCB design, PCB assembly, or cell phone skinning, while others have broader capabilities.

Since the shanzhai are small, they have to be efficient to maximize output. One shop of under 250 employees can churn out over 200,000 mobile phones per month with a high mix of products, sometimes producing runs as short as a few hundred units. Collectively, shanzhai in the Shenzhen area produced an estimated 20 million phones per month in 2009. That's an economy approaching a billion dollars a month. Most of those phones sell into third-world and emerging markets like India, Africa, Russia, and southeast Asia.

More Than Copycats

Significantly, the shanzhai's product portfolio includes more than just copycat phones. They innovate and riff on designs to make original products as well. These original phones integrate wacky features like 7.1 stereo sound, dual SIM cards, a functional cigarette holder, a high-zoom lens, or a built-in UV LED for counterfeit money detection.

The shanzhai do to hardware what the web did to mashup compilations. Mobile phones that are also toy Ferraris and watch-phone combos (complete with camera!) are good examples: they don't copy any single idea, but rather mix IP from multiple sources to create a new heterogeneous composition, such that the original source material is still distinctly recognizable in the final product. Also, like many web mashups, the result might seem nonsensical to a mass market (like the Ferrari phone) but is extremely relevant to a select long-tail market.

In a way, some shanzhai products are just ahead of their time; the watch-phones I saw, for example, predated smartwatches by several years.

Top: The front and back sides of a phone made to look like a pack of cigarettes. Bottom left: An Android-based smart watch, which unlike the Apple Watch includes a call-capable phone in the watch. Bottom right: A shanzhai-designed "baby iPhone," running Android, shown next to an Apple iPhone 6 for scale.

Community-Enforced IP Rules

The shanzhai also employ a concept called the *open BOM*: when one shanzhai builds something new, they share the bill of materials and other design documents with the others. If the product is based on an existing product, any improvements they make are also shared. These rules are policed by word of mouth within the community to the extent that if someone is found cheating, they are ostracized by the shanzhai ecosystem.

This system is viewed very positively in China. For example, I once heard a local say it was great that the shanzhai could not only clone an iPhone but also improve upon the original by giving the clone a user-replaceable battery. US law would call this activity illegal and infringing, but given the fecundity of mashup culture on the web, I can't help but wonder if hardware mashup isn't a bad thing. There's definitely a perception in the United States that if it's strange and it happens in China, it must be bad. This bias casts a long shadow over objective evaluation of a cultural phenomenon that could eventually be very relevant to the United States.

In a sense, the shanzhai are brethren of the classic Western notion of hacker-entrepreneurs, but with a distinctly Chinese twist. My personal favorite shanzhai story is about a chap who owns a three-story house that I am extraordinarily envious of. His bedroom is on top, the middle floor is a complete SMT manufacturing line, and the bottom floor is a retail outlet for the products produced a floor above and designed in his bedroom. Talk about a vertically integrated supply chain! Owning infrastructure like that would certainly disrupt the way I innovate. I could save on production costs, reduce my prototyping time, and aggressively turn inventory around, thereby reducing inventory capital requirements. And if my store were in a high-traffic urban location, I could also cut out the 20 to 50 percent minimum retail margin typically required by US retailers.

I have a theory that when the amount of knowledge and the scale of the markets in Shenzhen reach critical mass, the Chinese will stop being simply workers or copiers. They'll take control of their destinies and, ultimately, become innovation leaders. These stories about the shanzhai and their mashups are just the tip of an iceberg with the potential to change the way business is done—perhaps not in the United States, but certainly in that massive, untapped market often referred to as "the rest of the world."

THE $12 PHONE

Mashup cell phones demonstrate the shanzhai's innovation and willingness to experiment. But despite all the bells and whistles, those phones are quite affordable. One question you might ask, then, is how cheaply can you make a phone?

A short jaunt to the northeast corner of the Hua Qiang Bei electronics district brings you to the Mingtong Digital Mall. It's a four-story maze packed with tiny shops hawking all manner of quirky phones with features useful in economies that lack the infrastructure of consistent electricity or cable networks. For instance, some phones can run for a month thanks to comically oversized batteries. Others have analog TV tuners, integral hand-crank chargers, and multiple user profiles, enabling a family or small village to share a single phone.

During a visit to the Hua Qiang Bei district in 2013, I paid $12 for a complete phone, featuring quad-band GSM, Bluetooth, MP3 playback, an OLED display, and a keypad for the UI. It's nothing compared to a smartphone, but it's useful if you're going out and worried about your primary phone getting wet or stolen. And for a couple billion people, it may be the only phone they can afford.

Keep in mind this is the contract-free price. In countries that allow carriers to lock phones, such as the United States, phones are often given away or sold to buyers at a fraction of their cost in exchange for a subscription contract often worth several times the phone's value. The fact that I paid $12 over the counter for a contract-free, nonpromotional, unlocked, new-in-box phone with a charger, protective silicone sleeve, and cable means that the phone's production cost has to be somewhere below the retail price of $12. Otherwise, the phone's maker would be losing money. Rumors placed its cost below $10.

My simple but functional $12 phone

This is a really amazing price point. That's about the price of a large Domino's cheese pizza, or a decent glass of wine in an urban US restaurant. It's even cheap compared to an Arduino Uno. Admittedly, the comparison is a little unfair, but humor me and take a look at the specs for both, shown in Table 1.

Table 1: Comparing the $12 Phone with an Arduino

Spec	This phone	Arduino Uno
Price	$12	$29
CPU speed	260 MHz, 32-bit	16 MHz, 8-bit
RAM	8MiB	2.5kiB
Interfaces	USB, microSD, SIM	USB
Wireless	Quadband GSM, Bluetooth	—
Power	LiPo battery, includes adapter	External, no adapter
Display	Two-color OLED	—

How is it possible that this phone has better specs than an Arduino and costs less than half the price? I don't have the answers, but I'm trying to learn them. Tearing down the phone yielded a few hints.

Inside the $12 Phone

First, there are no screws in this phone. The whole case snaps together.

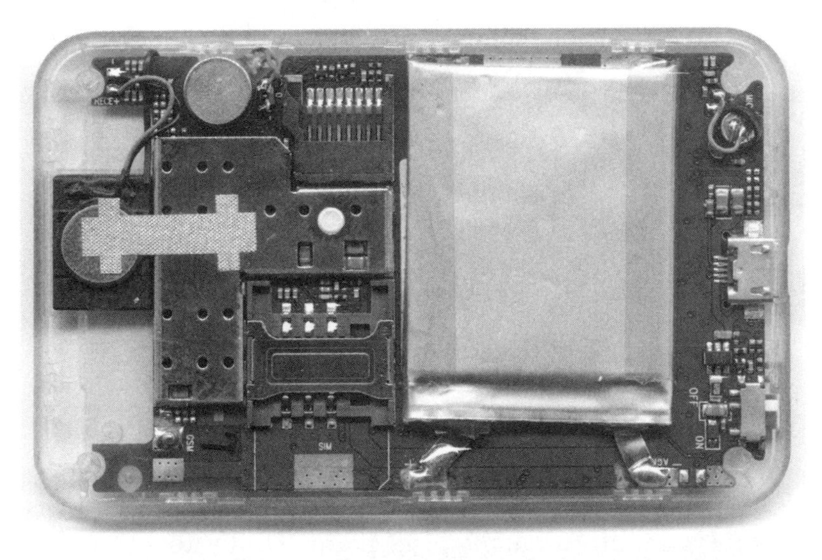

The back of the phone, after the cover is removed

There are (almost) no connectors on the inside. For shipping and storage, you get to flip a switch to hard-disconnect the battery. As best as I can tell, the battery also has no secondary protection circuit. Still, the phone features accoutrements such as a backlit keypad and decorative lights around the edge.

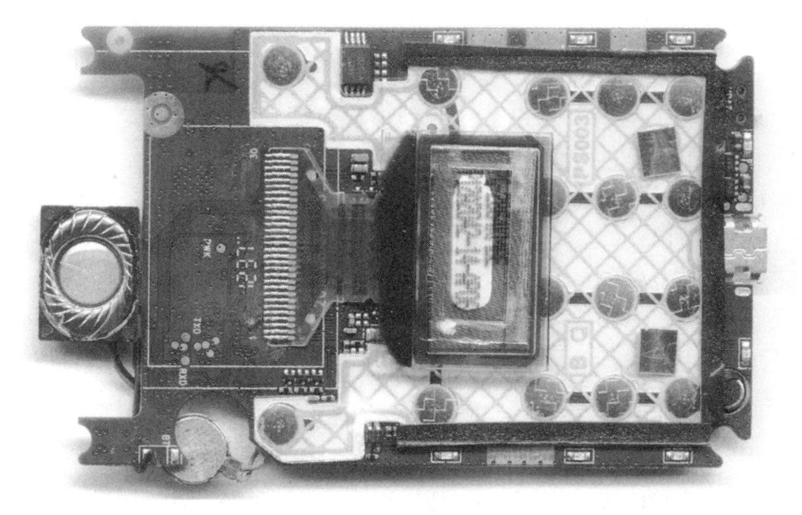

Everything from the display to the battery is soldered directly to the board.

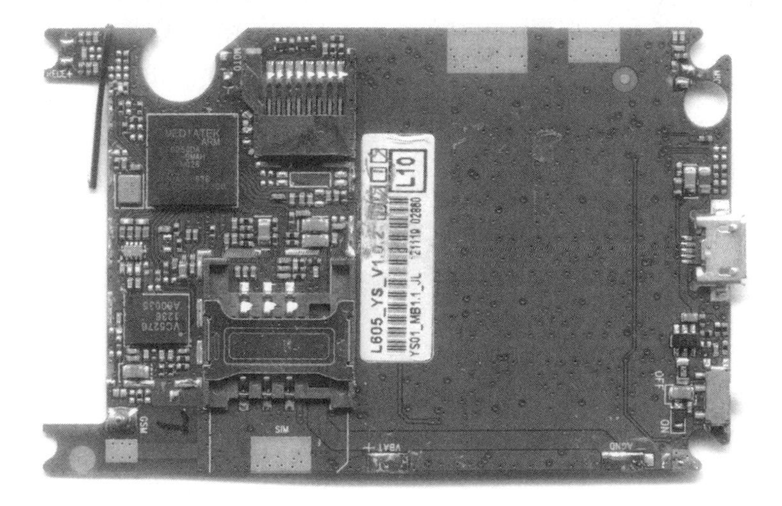

There are little decorative LEDs all over this PCB.

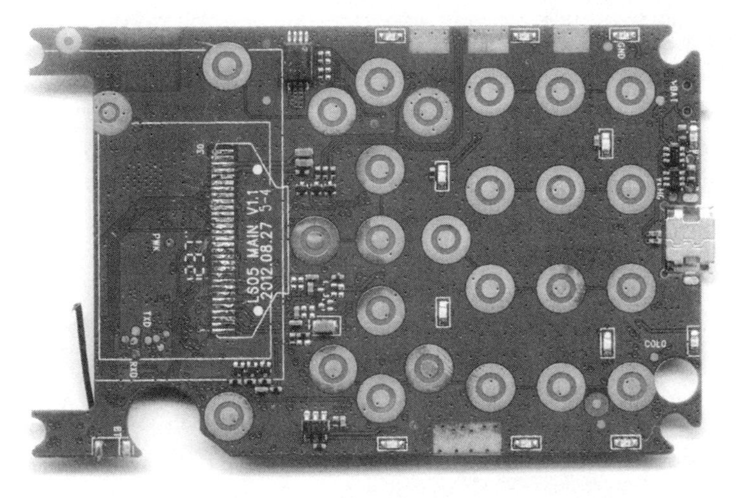

The Bluetooth antenna is the small length of wire on the bottom left.

The electronics consist of just two major ICs: the MediaTek MT6250DA and a Vanchip VC5276. The MT6250 is rumored to sell in volume for under $2. I was able to anecdotally confirm the price by buying a couple of pieces on cut tape from a retail broker for about $2.10 each.* That beats the best price I've ever been able to get on an ATMega of the types used in an Arduino. With price competition like this, Western firms are suing to protect ground: Vanchip got into a bit of a legal tussle with RF Micro, and MediaTek has been subject to a few lawsuits of its own.

Two MediaTek MT6250 ICs

* No, I will not broker these chips for you.

Of course, you can't just call up MediaTek and buy these chips. It's extremely difficult to engage with them "going through the front door" to do a design. However, if you know a bit of Chinese and the right websites, you can download schematics, board layouts, and software utilities for something similar to this phone, possibly with some different parts . . . for "free." *Free* is in quotes because you could obtain the source code but not the unambiguous legal right to use it, as the source code was distributed without the explicit legal consent of the copyright holders. But anyone unconcerned or unfamiliar with such legal frameworks could build versions of this phone, with minimal cash investment. It feels like open source, but it's not: it's a different kind of open ecosystem.

Introducing Gongkai

Welcome to the Galapagos of Chinese "open" source. I call it *gongkai* (公开), which is the Chinese transliteration of the English *open*, as applied to *open source*. There's a literal translation for *open source* into Chinese (*kaiyuan*), but the only similarity between gongkai practices and Western open source practices is that both allow you to download source code; the legal and cultural frameworks that enable such sharing couldn't be more different. It's like convergent evolution, where two species may exhibit similar traits, but the genes and ancestry are totally different.

Gongkai refers to the fact that copyrighted documents, sometimes labeled "confidential" and "proprietary," are made known to the public and shared overtly, but not necessarily according to the letter of the law. This copying isn't a one-way flow of value, as it would be in the case of copied movies or music. Rather, these documents are the knowledge base someone would need to build a phone using the copyright owner's chips, and sharing the documents promotes sales of their chips. There is ultimately a quid pro quo between the copyright holders and the copiers.

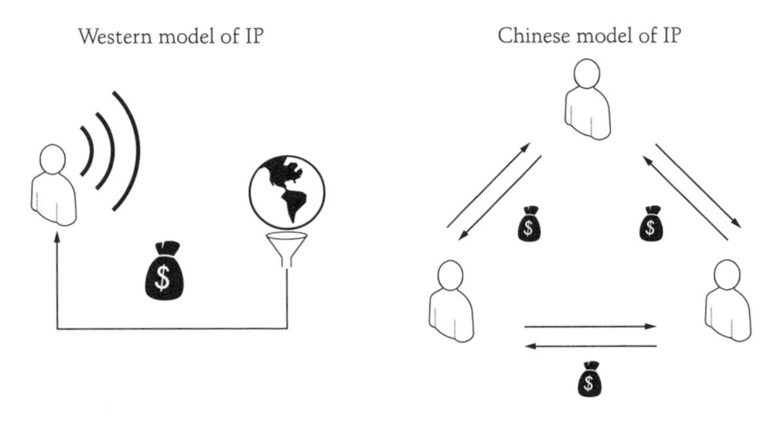

Comparing IP models. On the left, the Western "broadcast" model, with a single owner who controls and disseminates IP and is paid by society. On the right, the Chinese "network" model, where IP trades hands like a commodity, and payment is often in-kind or as favors.

This gray relationship between companies and entrepreneurs is just one manifestation of a much broader cultural gap between the East and the West. The West has a "broadcast" view of IP and ownership: good ideas and innovation are credited to a clearly specified set of authors or inventors, and society pays them a royalty for their initiative and good works. China has a "network" view of IP and ownership: one attains the far-reaching sight necessary to create good ideas and innovations by standing on the shoulders of others, and people trade these ideas as favors. In a system with such a loose attitude toward IP, sharing with the network is necessary, as tomorrow your friend could be standing on your shoulders, and you'll be looking to them for favors.

In the West, however, rule of law enables IP to be amassed over a long period of time, creating impenetrable monopoly positions. That's good for the guys on top but tough for upstarts, causing a situation like the modern Western cell phone market. Companies like Apple and Google build amazing phones of outstanding quality, and startups can only hope to build an "appcessory" for their ecosystem.

I've reviewed business plans for over 100 hardware startups, and the foundations for most are overpriced chipsets built with antiquated process technologies. I'm no exception to this rule; the Novena uses a Freescale (now NXP after an acquisition) i.MX6 processor, which was neither the cheapest nor the fastest chip on the market when I designed the laptop. But it's a chip with two crucial qualities: anyone can freely download almost complete documentation for it, and anyone can buy it on Digi-Key.

Scarce documentation and supply for cutting-edge technology force Western hardware entrepreneurs to look primarily at Arduino, Beaglebone, and Raspberry Pi as starting points for their good ideas. Chinese entrepreneurs, on the other hand, churn out new phones at an almost alarming pace.

Every object pictured here is a phone.

Phone models change on a seasonal basis. Entrepreneurs experiment all the time, integrating wacky features into phones, such as cigarette lighters, extra-large battery packs (to charge a second phone), huge buttons (for the visually impaired), call-home buttons only (to give to children for emergencies), watch form factors, and so on. This works because small teams of engineers can obtain complete design packages for working phones—including the case, board, and firmware—allowing them to fork the design and focus only on changing the pieces they really care about.

As a hardware engineer, I want that.

I want to be able to fork existing cell phone designs. I saw the $12 phone, and I, too, wanted to use a 364 MHz 32-bit microcontroller with megabytes of integrated RAM and dozens of peripherals that costs $3 in single quantities. The Arduino Uno's ATMega microcontroller, a 16 MHz 8-bit microcontroller with a few kilobytes of RAM and a smattering of peripherals, pales in comparison yet costs twice as much, at $6.

From Gongkai to Open Source

So, I decided to take my study of the phone one step further from a teardown, and attempt to make my own version—in the style of the shanzhai, but interpreted through Western eyes. That's how Sean "xobs" Cross and I started a project we dubbed *Fernvale*. Sean has been my adventure partner on dozens of projects since we first met at Chumby, where I recognized his talent as a firmware engineer when he showed me how he ported Quake to chumby in his spare time. Sean has always marched to the beat of his own drum. Born in Germany to American parents, he studied cognitive science in college, and prior to working at Chumby, he spent six months wandering New Zealand and Australia, searching for adventure and work. At Chumby, he was easy to spot, thanks to his ponytail and kilt (actually, a Utilikilt).

After Chumby went out of business, Sean and I found ourselves washed up on the shores of Singapore, where I started a boutique hardware consulting firm called Sutajio Ko-Usagi, which is *bunniestudios* translated to Japanese and then romanized into English characters. Sean's virtuoso coding abilities have been an excellent complement to my hardware design skills, and since then, we've completed several significant open source projects.

We figured at first we should at least try to go "through the front door" and inquire directly with the chipmakers about what it might take to get a proper Western-licensed embedded development kit (EDK) for the chips used in these shanzhai phones. Our inquiries were met with a cold shoulder. I was told the volumes for our little experiment were too small, or we'd have to enter minimum purchase agreements backed by a prohibitive cash deposit in the hundreds of thousands of dollars.

Even for people who jump through such hoops, these EDKs don't include all the reference material the Chinese get to play with. The datasheets are incomplete, and you're forced to use the companies' proprietary OS ports. It feels like a case of the nice guys finishing last. Could we find a way to get ahead yet still play nice?

Engineers Have Rights, Too

Thus, Fernvale had two halves: the technical task of reverse engineering and re-engineering the phone and the legal task of creating a general methodology for absorbing gongkai IP into the Western ecosystem. I'll recount the technical task in Chapter 9 and focus on the legal task for the remainder of this chapter.

After some research into the legal frameworks and challenges, I believed I'd found a path to repatriate some of the IP from gongkai into proper open source. I must, however, give a disclaimer: I'm not a lawyer. I'll tell you my beliefs, but don't construe them as legal advice.*

* I've often wondered why the "I am not a lawyer" disclaimer is necessary. It was explained to me that even the appearance of dispensing legal advice without the disclaimer can make me guilty of practicing law without a proper license. I could also be held accountable for bad decisions made by people who construe the opinions as legal advice.

My basic idea with Fernvale was to exercise the right to reverse engineer in a careful, educated fashion to increase the likelihood that, if push came to shove, the courts would agree with my actions. But I also feel that shying away from reverse engineering simply because it's controversial is a slippery slope: to have your rights, you must exercise them. If women didn't vote and black people sat in the back of the bus because they were afraid of controversy, the United States would still be segregated and without universal suffrage. Although reverse engineering is a trivial issue compared to racial equality and universal suffrage, the precedent is clear: in order to have rights, you must be bold enough to stand up and assert them.

DEALING WITH PATENTS AND OTHER LAWS

Open source has two broad categories of IP issues to deal with: patents and copyrights. Patents present complex issues, and it seems the most practical approach is to essentially punt on the issue. For instance, nobody, as far as I know, checks their Linux commits for patent infringement before upstreaming them, and in fact, many corporations have similar policies at the engineering level.

Why? Determining which patents apply and if a product infringes takes a huge amount of resources. Even after expending those resources, you can't be 100 percent sure. Further, becoming very familiar with the body of patents amplifies the possibility that any infringement is willful, thus tripling damages. Finally, it's not even clear where the liability for infringement lies, particularly in an open source context.

Thus, Sean and I did our best not to infringe with Fernvale, but we couldn't be 100 percent sure that no one would allege infringement. However, we did apply a license to our work that includes a "poison pill" clause for patent holders who might attempt to litigate. Poison pills make the entire body of open source work unavailable to any party who files a lawsuit alleging infringement of any part against any entity.*

For copyrights, the issue is also extremely complex. The Coders' Rights Project from the Electronic Frontier Foundation (EFF) has a Reverse Engineering FAQ† that's a good read if you really want to dig into the issues. To sum it up, courts have found that reverse engineering to understand the ideas embedded in code and to achieve interoperability is fair use. As a result, anyone likely has the right to study the gongkai-style IP, understand it, produce a new work, and apply a Western-style Open IP license to it.

However, before I could attack the copyright issues for Fernvale, I had to make sure we wouldn't bump into other laws that could impede our fair use rights. First, there's the Digital Millennium Copyright Act (DMCA). The DMCA makes circumventing any encryption designed to enforce a copyright basically illegal, with only a few poorly tested exemptions allowed. Since none of the files or binaries Sean and I downloaded were encrypted or had access controlled by any technological measure, we didn't have to do any circumvention. No circumvention, no DMCA problem.

All the files we obtained came from searches linking to public servers, so there would be no Computer Fraud and Abuse Act (CFAA) problems. None of the devices we used in the work came with shrink-wraps, click-throughs, or other end-user license agreements (EULAs), terms of use, or other agreements that could waive our rights.

---

* Specifically, Apache 2.0, section 3 reads, "Grant of Patent License. . . . If You institute patent litigation against any entity (including a cross-claim or counterclaim in a lawsuit) alleging that the Work or a Contribution incorporated within the Work constitutes direct or contributory patent infringement, then any patent licenses granted to You under this License for that Work shall terminate as of the date such litigation is filed."

† *https://www.eff.org/issues/coders/reverse-engineering-faq/*

DEALING WITH COPYRIGHTS

With the DMCA, CFAA, and EULA concerns set aside, we were finally able to address the core issue: what to do about copyrights.

The cornerstone of our methodology hinged on decisions rendered on several occasions by courts stating that facts are not copyrightable. For example, Justice O'Connor wrote the following in *Feist Publications, Inc. v. Rural Telephone Service Co., Inc.* (449 U.S. 340, 345, 349 (1991):*

> Common sense tells us that 100 uncopyrightable facts do not magically change their status when gathered together in one place. . . . The key to resolving the tension lies in understanding why facts are not copyrightable: The sine qua non of copyright is originality.

And:

> Notwithstanding a valid copyright, a subsequent compiler remains free to use the facts contained in another's publication to aid in preparing a competing work, so long as the competing work does not feature the same selection and arrangement.

Based on this opinion, anyone has the right to extract facts from proprietary documentation and carefully re-express those facts in their own selection and arrangement. Just as the facts that "John Doe's phone number is 555-1212" and "John Doe's address is 10 Main St." are not copyrightable, facts such as "The interrupt controller's base address is 0xA0060000" and "Bit 1 controls status reporting of the LCD" aren't copyrightable, either. Sean and I extracted such facts from datasheets and re-expressed them in our own header files where, as the legal owners of newly created expressive speech, we applied a proper open source license of our choice.

MAKING A PROGRAMMING LANGUAGE

But the situation was further complicated by hardware blocks we had absolutely no documentation for. In some cases, we couldn't even learn what a block's registers meant or how the

* See also Sony Computer Entertainment, Inc. v. Connectix Corp., 203 F. 3d 596, 606 (9th Cir. 2000) and Sega Enterprises Ltd. v. Accolade, Inc., 977 F.2d 1510, 1522-23 (9th Cir. 1992).

blocks functioned from a datasheet. For these blocks, we isolated and extracted the code responsible for initializing their state. We then reduced this code into a list of address and data pairs, and expressed it in a custom scripting language we called *scriptic*. We invented our own language to avoid subconscious plagiarism—it's too easy to read one piece of code and, from memory, code something almost exactly the same. By transforming the code into a new language, we were forced to consider the facts presented and express them in an original arrangement.

Scriptic is basically a set of assembler macros, and the syntax is very simple. Here is an example of a scriptic script:

```
#include "scriptic.h"
#include "fernvale-pll.h"

sc_new "set_plls", 1, 0, 0

  sc_write16 0, 0, PLL_CTRL_CON2
  sc_write16 0, 0, PLL_CTRL_CON3
  sc_write16 0, 0, PLL_CTRL_CON0
  sc_usleep 1

  sc_write16 1, 1, PLL_CTRL_UPLL_CON0
  sc_write16 0x1840, 0, PLL_CTRL_EPLL_CON0
  sc_write16 0x100, 0x100, PLL_CTRL_EPLL_CON1
  sc_write16 1, 0, PLL_CTRL_MDDS_CON0
  sc_write16 1, 1, PLL_CTRL_MPLL_CON0
  sc_usleep 1

  sc_write16 1, 0, PLL_CTRL_EDDS_CON0
  sc_write16 1, 1, PLL_CTRL_EPLL_CON0
  sc_usleep 1

  sc_write16 0x4000, 0x4000, PLL_CTRL_CLK_CONDB
  sc_usleep 1

  sc_write32 0x8048, 0, PLL_CTRL_CLK_CONDC
  /* Run the SPI clock at 104 MHz */
  sc_write32 0xd002, 0, PLL_CTRL_CLK_CONDH
  sc_write32 0xb6a0, 0, PLL_CTRL_CLK_CONDC
  sc_end
```

This script initializes the Phase Locked Loop (PLL, a circuit for generating clock waveforms) on the target chip for Fernvale, the MediaTek MT6260. To contrast, here are the first few lines of the code snippet from which that scriptic code was derived:

```
// enable HW mode TOPSM control and clock CG of PLL control

*PLL_PLL_CON2 = 0x0000; // 0xA0170048, bit 12, 10 and 8 set to 0
                        // to enable TOPSM control
                        // bit 4, 2 and 0 set to 0 to enable
                        // clock CG of PLL control
*PLL_PLL_CON3 = 0x0000; // 0xA017004C, bit 12 set to 0 to enable
                        // TOPSM control

// enable delay control
*PLL_PLLTD_CON0= 0x0000; // 0x A0170700, bit 0 set to 0 to
                         // enable delay control

// wait for 3us for TOPSM and delay (HW) control signal stable
for(i = 0 ; i < loop_1us*3 ; i++);

// enable and reset UPLL
reg_val = *PLL_UPLL_CON0;
reg_val |= 0x0001;
*PLL_UPLL_CON0  = reg_val; // 0xA0170140, bit 0 set to 1 to
                           // enable UPLL and
                           // generate reset of UPLL
```

The original code actually goes on for pages and pages, and even this snippet is surrounded by conditional statements, which we culled as they were irrelevant to initializing the PLL correctly.

Knowledge of our rights, a pool of documentation to extract facts from, and scriptic were tools in our armory. With them, Sean and I derived sufficient functionality for our Fernvale project to eventually boot a small, BSD-licensed, real-time operating system (RTOS) known as NuttX, running on our own custom hardware. I'll go more into the gory details of how we did that in Chapter 9.

CLOSING THOUGHTS

Rights atrophy and get squeezed out by competing interests if they aren't vigorously exercised. Sean and I did Fernvale because we think it's imperative to exercise our fair use rights to reverse engineer and create interoperable, open source solutions. For decades, engineers have sat on the sidelines and seen ever more expansive patent and copyright laws shrink their latitude to learn freely and to innovate. I'm sad that the formative tinkering I did as a child is no longer a legal option for the next generation of engineers.

The rise of the shanzhai and their amazing capabilities is a wake-up call. I see it as evidence that a permissive IP environment spurs innovation, especially at the grassroots level. If more engineers learn their fair use rights and exercise them vigorously and deliberately, perhaps this can catalyze a larger and much-needed reform of the patent and copyright system. Our Fernvale project is hopefully just a signpost pointing the way for much bigger efforts to bridge the gap between the gongkai and open source communities.

Being able to cherry-pick the positive aspects of gongkai into the Western IP ecosystem is an important tool. Rule of law has its place, and an overly permissive system has its own problems. The next chapter explores some of the negative consequences of an overly permissive IP ecosystem: fake and counterfeit goods.

5. fake goods

The gongkai system fosters an amazing amount of innovation in China, and the shanzhai can make interesting original products, like the cell phones I showed you in Chapter 4. That said, China does produce plenty of fake electronic goods, and they aren't all knockoff iPhones. Clever counterfeiters can produce fake integrated circuits, including microSD cards and even FPGAs.

WELL-EXECUTED COUNTERFEIT CHIPS

For instance, in 2007 (while I was still working with Chumby) I encountered some counterfeit chips so well executed that I couldn't be certain they were fake without investigating.

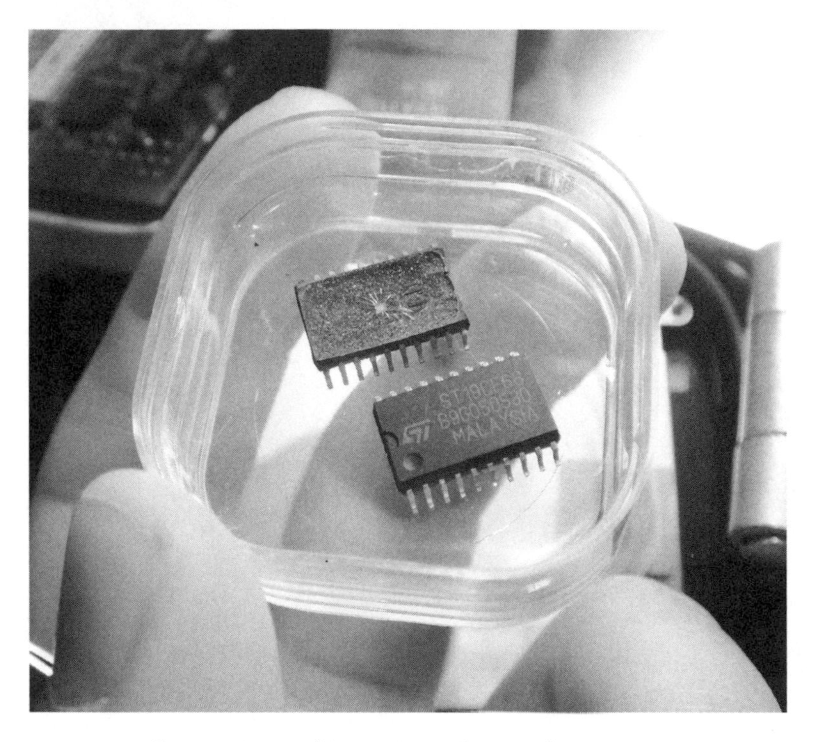

Two suspicious chip specimens from an Asian source

The chips claimed to be ST19CF68s, a chip made by STMicroelectronics and described on its datasheet as a "CMOS MCU Based Safeguard Smartcard I/O with Modular Arithmetic Processor." ST19CF68 chips are normally sold prepackaged in *smartcard* (for example, the chip on the front of a credit card) or *diced wafer* (a silicon wafer that's been diced into individual chips, but with no other package around it) format, but curiously, these were SOIC-20 packaged devices. To find out the reason for the odd package choice, I dissolved the black epoxy packaging off the top of one chip to decapsulate it so I could inspect the silicon on the inside using a microscope.

The die inside the package was much too small and simple for a complex microcontroller unit (MCU) matching the description of the ST19CF68. The pattern of gold-colored rectangles tiled across the chip was too coarse; I could make out individual transistors at low zoom with an optical microscope. The size of these features is referred to as the chip's *process geometry*. The process geometry of a smartcard would typically trail a cutting-edge CPU by at most three or four generations, making transistors very difficult to resolve even at the highest levels of zoom.

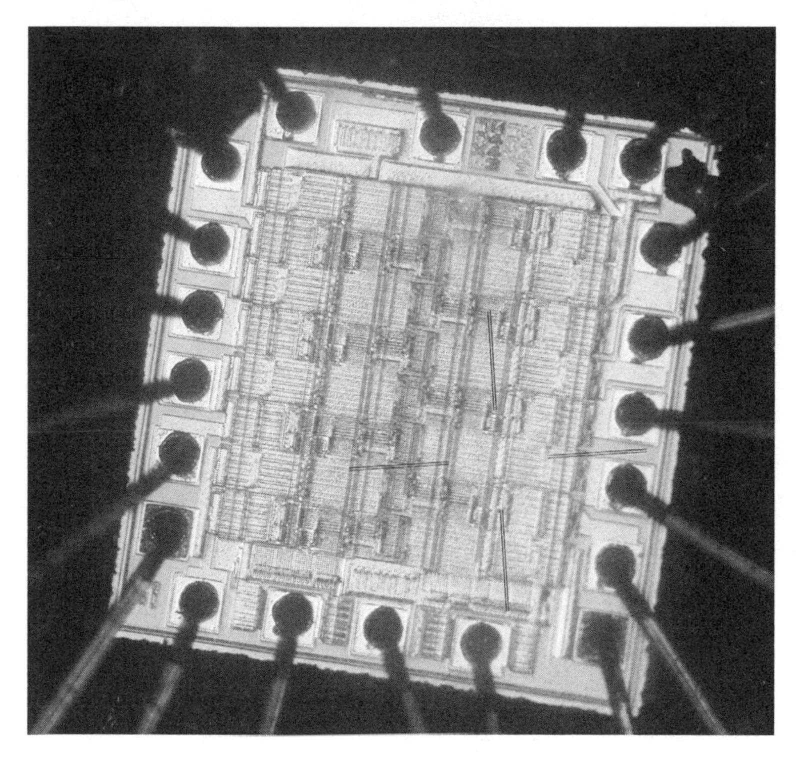

The silicon inside the fake ST19CF68

Along with the unexpectedly coarse process geometry, why did this part have 20 bondable pads and 20 pins when, according to the datasheet, it should have only 8 pads? Zooming in a bit on the die revealed some interesting details.

The chip manufacturer and copyright date

The chip wasn't made by STMicroelectronics after all! The label on the silicon said *FSC*, indicating it was made by Fairchild Semiconductor. Of course, then I had to check the part label on the silicon, too.

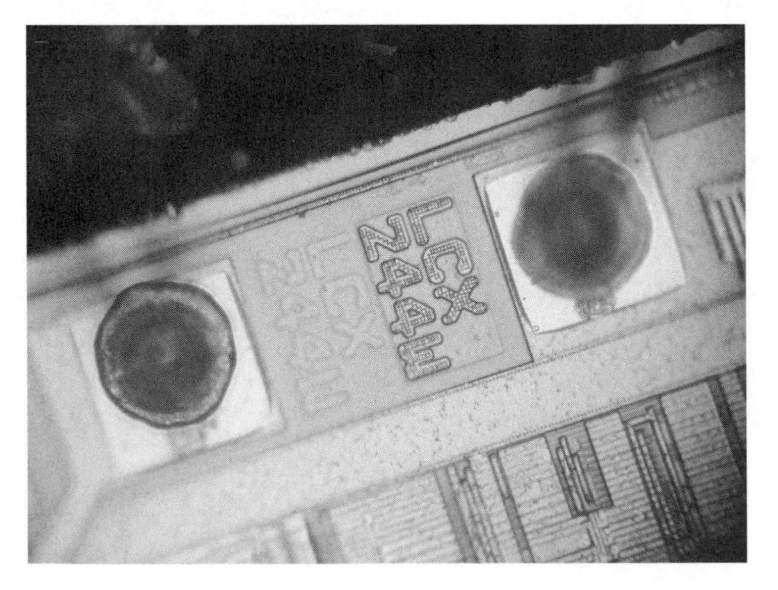

Discovering the true part number

The die within that chip turned out to be a Fairchild 74LCX244, which is a "Low Voltage Buffer/Line Driver with 5V Tolerant Inputs and Outputs." The 74LCX244 is a much cheaper piece of silicon than the ST19CF68 the package supposedly contained.

Of course, the mismatched pin count was suspicious, but manufacturers have been known to put chips in larger packages, especially during early runs of the chip before it has been size-optimized. The thing that really got me was the convincing quality of the package and the markings.

Normally, remarked or fake chips look cheesier than this one. The original chips are sanded down or painted over to remove the previous markings, and the new marking is typically applied with silkscreened paint.

But these chips showed no evidence of remarking at all. The markings are of first-run quality: someone acquired unmarked blanks of the 74LCX244 chip and programmed a production laser engraver to put high-quality fake markings on an otherwise virgin package. They even got the proportions of the *ST* logo exactly right.

A close-up of the outside of the fake ST19CF68

The quality difference between a remarked chip and first-run marking is like the quality difference between spray paint used to hide a scratch on a car and the car's original, factory-fresh paint job. This chip definitely had the "new car" look.

This discovery left me with a lot of unanswered questions. How did someone acquire unmarked Fairchild silicon? Was the person an insider, or did Fairchild sloppily throw away unmarked reject chips without grinding them up or clipping off leads so they couldn't be picked out of a dumpster and resold? The laser-marking machine used to make those markings wasn't a cheap desktop engraver, either; it had to be a high-power raster engraver, and the artwork was spot-on.

I still find it hard to believe those fake chips were made and sold, but maybe I shouldn't. I've seen brazen remarking of dual inline memory modules (DIMMs, the memory used in personal computers) in the SEG Electronics Market, and many counterfeiters at the market openly display their arsenal of professional-quality thermal transfer label printers and hologram sticker blanks.

If fakes of this quality become more common, they could present a problem for the supply chain. Clearly, whoever made the counterfeit ST19CF68 can fake just about any chip, and the fakes are gradually appearing on the US market. Resellers, especially distributors that specialize in buying excess manufacturer inventory, implicitly trust the markings on a chip.

I don't think chipmakers will put anticounterfeiting measures on chip markings, but the quality of these fakes definitely made me wary when I discovered them, and it still does. Not all fakes get spotted before they're used, and fake components pose problems in any project where they appear.

COUNTERFEIT CHIPS IN US MILITARY HARDWARE

Counterfeit chips can be particularly problematic when they find their way into military projects. The US military has a unique problem: it's one of the biggest and wealthiest buyers of really old parts because military designs have shelf lives of decades. Like anything else, the older a part is, the harder it is to find, and sometimes contractors are sold fakes. For example, a 2011 Senate hearing report revealed that some parts used in the P-8 Poseidon (a plane the US Navy commissioned from Boeing) were, as an article from the Defense Tech website put it, "badly refurbished," causing a key system to fail.

The US government attempted to reduce fakes in its supply chain with Amendment 1092 to the National Defense Authorization Act for Fiscal Year 2012 (H.R. 1540). The amendment is a well-intentioned but misguided provision outlining measures designed to reduce the prevalence of counterfeit chips in the US military supply chain.

Even before Amendment 1092 was put on the table, the Defense Authorization Act drew flak for a provision that authorizes the US military to detain US citizens indefinitely without trial. It also rather ironically requires an assessment of the US federal debt owed China as a potential "national security risk" (section 1225 of H.R. 1540).

Under the anticounterfeit amendment, first-time offenders can receive a $5 million fine and 20-year prison sentence for individuals, or a $15 million fine for corporations—a penalty comparable to that of trafficking cocaine.* While the amendment explicitly defines *counterfeit* to include refurbished parts represented as new, the wording is regrettably vague on whether you must be willfully trafficking such goods to also be liable for such a stiff penalty.

---

* See Sec 2320 (b) at *https://www.govtrack.us/congress/bills/112/hr1540/text.*

If you took a dirty but legitimately minted coin and washed it so that it looked mint condition, nobody would accuse you of counterfeiting. Yet this amendment puts a 20-year, $5 million penalty not only on the act of counterfeiting chips destined for military use but also potentially on the unwitting distribution of refurbished chips that you putatively bought as new. Unfortunately, in many cases an electronic part can be used for years with no sign of external wear.

The amendment also has a provision to create an "inspection program":

> (b) Inspection of Imported Electronic Parts —
>
> (1) . . . the Secretary of Homeland Security shall establish a program of enhanced inspection by U.S. Customs and Border patrol of electronic parts imported from any country that has been determined by the Secretary of Defense to have been a significant source of counterfeit electronic parts . . .

Inspecting fruits and vegetables as they enter the country for pests and other problems makes sense, but requiring customs officers to become experts in detecting fake electronic components seems misguided. Burdening vendors with detecting fakes when there are such high penalties for failure is also misguided, given how easy it is for forgers to create high-quality counterfeits.

Types of Counterfeit Parts

To better understand the magnitude of the chip counterfeiting problem, let's look at how fakes are made. The fake chips I've seen fall into the following broad categories.

EXTERNAL MIMICRY

The most trivial counterfeit chips are simply empty plastic packages with authentic-looking top marks, or remarked parts that share only physical traits with the authentic parts. For example, a simple transistor-transistor logic (TTL) chip might be placed inside the same package, with identical markings, as an expensive microcontroller.

I consider external mimicry trivial because fakes produced this way are easy to detect in a factory test. At worst, you're sold a mixture of mostly authentic parts with a few counterfeits blended in so that testing just one part out of a tube or reel isn't good enough to catch the issue. But most products employ 100 percent testing at the system level, so typically the problem is discovered before anything leaves the factory.

REFURBISHED PARTS

Counterfeits don't technically have to be fake at all, though. Refurbished parts are authentic chips that are desoldered from e-waste and reprocessed to look new. They're very difficult to spot since the chip is in fact authentic, and a skilled refurbisher can produce stunningly new-looking chips that only isotopic or elemental analysis could identify as used.

This category also includes parts that are "new" in the sense that they've never been soldered onto a board but have been stored improperly, perhaps in a humid environment. Such chips should be scrapped but are sometimes stuck in a fresh foil pack with a more recent date code, and sold as new.

REBINNED PARTS

Counterfeiters sometimes remark authentic parts that have never been used (and so can be classified as new) as a better version of an otherwise identical part. A classic example is grinding and remarking CPUs with a higher speed grade, or more trivially, marking parts that contain lead as RoHS-compliant.

But rebinning can get more sophisticated. Vendors may reverse engineer and reprogram the fuse codes inside the remarked chip so that the chip's electronic records actually match the faked markings on top. Vendors have also been known to hack flash drive firmware so that a host operating system will perceive a small memory as much larger. Such hacks even go so far as to "loop" memory so that writes beyond the device capacity appear to succeed, thus requiring a time-consuming full readback and comparison of the written data to detect the issue.

GHOST-SHIFT PARTS

Some fakes are created on the exact same fabrication facility as authentic parts; they're run very late at night by rogue employees without the manufacturer's authorization and never logged on the books. These unlogged production runs are called *ghost shifts*. It's like an employee in a mint striking extra coins after-hours. Ghost-shift parts are often assigned a lot code identical to a legitimate run, but certain testing steps are skipped.

Ghost shifts often use marginal material left over from the genuine product that would normally be disposed of but was intercepted on the way to the grinder. As a result, the markings and characteristics of the material often look absolutely authentic. These fakes can be extremely hard to detect.

FACTORY SCRAP

Factory rejects and prototype runs can be recovered from the scrap heap for a small bribe, given authentic markings, and resold as new. To avoid detection, workers often replace the salvaged scrap with physically identical dummy packages, thus foiling attempts to audit the scrap trail. This practice of replacing salvageable scrap with dummy fakes helps drive the market for the trivial "external mimicry" fakes. The existence of an industry that supplies low-quality fakes to dodge audits that would otherwise prohibit high-quality fakes gives you an idea of how sophisticated and mature the counterfeiting industry has become.

SECOND-SOURCING GONE BAD

Second-sourcing is a standard industry practice where competitors create pin-compatible replacements for popular products to drive price competition and strengthen the supply chain against events like natural disasters. The practice goes bad when inferior parts are remarked with the logos of premium brands.

High-value but functionally simple discrete analog chips such as power regulators are particularly vulnerable to this problem. Premium US-branded power regulators sometimes fetch a price 10 times higher than drop-in Asian-branded substitutes. However, the Asian-branded parts are notorious for spotty quality, cut corners, and poor parametric performance. Clearly, there is ample opportunity for counterfeiters to make a lot of money by buying unmarked chips from the second-source fab and remarking them with authentic-looking top marks of premium US brands. In some cases, there are no inexpensive or fast tests to detect these fakes, short of decapsulating the chip and comparing mask patterns and cross-sections, as I did for the ST19CF68.

Fakes and US Military Designs

The variety of counterfeiting methods available, combined with the fact that many commodity parts have production cycles of only a few years, presents a big problem for institutions like the US military, where design lifetimes are often measured in decades. It's like asking someone to build a NeXTcube* motherboard today using only certifiably new parts, with no secondhand or refurbished parts allowed. I don't think it's possible.

The impossibility of this situation may sometimes make military contractors complicit in the consumption of counterfeit parts to bad effect. In the P-8 Poseidon case, people were quick to point fingers at China, but a poor refurbishing

* Remember that one? The NeXTcube was a computer released in 1990 by Steve Jobs's company, NeXT.

job is probably detectable with a simple visual inspection. Maybe part of the problem is that a subcontractor was lax in checking incoming stock—or perhaps looking the other way. If those parts were the last of their kind in the world, what else could be done?

My guess is that the stocks of any distributor in the second-hand electronics business are already flooded with undetected counterfeits. Remember, only the bad fakes are ever caught, and chip packaging was not designed with anticounterfeiting measures in mind. While all gray-market parts are suspect, that's not necessarily a bad thing.

Gray markets play an essential role in the electronics ecosystem; using them is a calculated, but sometimes unavoidable, risk. In fact, many traders in the gray market are very upfront about their goods being recycled. Many even post signs on their stalls advertising this fact. However, these signs are written in Chinese. In that case, whose fault is it—the seller for selling recycled goods, or the buyer for not being able to read the sign?

Anticounterfeit Measures

The counterfeit chip situation is a mess, but some simple measures could fix it.

PHYSICAL IDENTIFIERS

Embedding anticounterfeit measures in chips approved for military use is one option. For chips larger than 1 cm wide, a unique 2D barcode could be laser-engraved by equipment relatively common in chip packaging facilities. Despite a tiny footprint, the codes would be backed with a guarantee of 100 percent uniqueness. Such techniques are effective in biotech, where systems like Matrix 2D track disposable sample tubes in biology labs.

Another potential solution is to mix a UV dye into the component's epoxy that changes fluorescence properties upon exposure to *reflow* temperatures—a consistent set of well-defined temperatures at which solder melts. This makes it impossible to recondition the chip to a "new" state after it's been soldered down the first time. If the dye is distributed through the entire package body, it will be impossible to remove with surface grinding alone.

CHANGING HOW E-WASTE IS HANDLED

Managing e-waste more effectively would also alleviate the counterfeit problem. E-waste is harvested in bulk for used parts. Crudely desoldered MSM-series chips—the brains of many Android smartphones, made by Qualcomm and marketed under the brand name of Snapdragon—are purchasable by the pound, at around 10 cents per chip. Counterfeiters clean up the chips, *reball* (that is, add new solder balls, for ball-grid array packages) and sometimes remark them, put them into tapes and reels, and sell them as brand-new, commanding a markup 10 times the original purchase price. A single batch of refurbished chips can net thousands of dollars, making the practice a compelling source of income for skilled workers who would otherwise earn $200 per month in a factory doing exactly the same thing.* (Factories are typically authorized to recover chips off of defective boards or consumer returns that can't be repaired.)

If the United States stopped shipping e-waste overseas for disposal, or at least ground up the parts before shipping them, then the supply for refurbished chip markets would decrease. Domestic e-waste processing would also create more jobs, a resource as valuable as gold.

On the other hand, I think component-level recycling is quite good for the environment and the human ecosystem in the long term. Most electronic parts will function perfectly for

* This was the salary rate in the mid-2000s; due to wage inflation since then, it's risen to around $1,000 per month, but refurbishing chips is still more lucrative.

years beyond a consumer's trash bin, and emerging economies create technology-hungry markets that can't afford new parts purchased on the primary market.

KEEPING A RESERVE OF AUTHENTIC PARTS

A final option to ensure trustworthiness for critical military hardware could be to establish a strategic reserve of parts. A production run of military planes is limited to perhaps hundreds of units, a small volume compared to consumer electronics production runs. I imagine the lifetime demand of a part, including replacements, is limited to tens of thousands of units. Physically, then, a parts reserve isn't unmanageable: 10,000 chips will fit inside a large shoebox.

Financially, I estimate purchasing a reserve of raw replacement components for critical avionics systems would add only a fraction of a percent to the cost of an airplane. This could even lead to long-term savings, as manufacturers can achieve greater scale efficiency if they run one large batch all at once.

Obviously, anticounterfeit measures would be incredibly useful in civilian projects, too. I have sympathy for anyone who has to deal with counterfeit parts, as I myself have been burned on several occasions. Here's a tale of a particularly annoying issue I ran into during my work on the chumby One.

FAKE MICROSD CARDS

In December 2009, in the middle of the chumby One's production run, I set out on a forensic investigation to find the truth behind some irregular Kingston memory cards. The factory called to tell me that SMT yield dropped dramatically on one lot of chumby Ones, so I drove over to see what I could do to fix the problem. After poking and prodding at some chumby Ones, I realized that all failing units had Kingston microSD cards from a particular lot code. I had the factory pull the entire lot of microSD cards from the line and rework the units that had these cards loaded. After swapping the cards, yield returned to normal.

The story should have ended there. In this situation, I'd usually get a return merchandise authorization (RMA) from the manufacturer for the defective parts, exchange the lot for parts that work, and move on. But I had a couple of problems.

First, Kingston wouldn't take the cards back, because we programmed them. Second, there were a lot of defective cards (about 1,000 altogether, and chumby was already deeply back-ordered), and memory cards aren't cheap. This type of memory card cost around $4 or $5 at the time, leaving a few thousand dollars in scrap if we couldn't get them exchanged. Chumby couldn't afford to sneeze at a few kilobucks, so I kicked into forensics mode.

Visible Differences

Irregular external markings were the first suspicious feature I noticed about the defective Kingston cards.

An irregular microSD card (left) and a normal card (right). The arrows and circles show suspicious differences.

The strangest physical difference was that the lot code on the irregular card was silkscreened with the same stencil as the main logo. Silkscreening a lot code isn't unusual, but typically, the manufacturer won't use the same stencil for the lot code and the logo. There should be some variance in the coloration, font, or alignment of the lot code from the rest of the text. The entire batch of irregular cards also had the same lot code (N0214-001.A00LF). Typically, the lot code changes at least every couple hundred cards. Contrast the irregular card with the normal card, which is laser-marked. The normal cards' lot codes varied with every tray of 96 units.

The second strange feature was subtler and perhaps not damning: an irregularity in the microSD logo. Brand-name vendors like Kingston are very picky about the accuracy of their logos: SanDisk cards have a broken *D*, but Kingston cards sold in the United States almost universally use a solid *D*.

Investigating the Cards

Oddities in the external markings were just the start. When I read the electronic card ID data on the two cards (by checking */sys* entries in Linux), this is what I found in the irregular card:

```
cid:41343253443247422000000960400049
csd:002600325b5a83a9e6bbff8016800095
date:00/2000
fwrev:0x0
hwrev:0x2
manfid:0x000041
name:SD2GB
oemid:0x3432
scr:0225000000000000
serial:0x00000960
```

And this is what I found in the normal card:

```
cid:02544d5341303247049c62cae60099dd
csd:002e00325b5aa3a9ffffff800a80003b
date:09/2009
fwrev:0x4
hwrev:0x0
manfid:0x000002
```

```
name:SA02G
oemid:0x544d
scr:0225800001000000
serial:0x9c62cae6
```

First, notice the date code on the irregular card. Dates are counted as the offset from 00/2000 in the CID field, so a value of 00/2000 means the manufacturer didn't bother to assign a date. Furthermore, in the year 2000, 2GB microSD cards didn't even exist. Also, the serial number on the defective card is very low: in decimal, 0x960 is 2,400. Other cards in the irregular batch had similarly low serial numbers, in the hundreds or thousands.

For a popular product like a microSD card, the chance of getting the very first units out of a factory is pretty remote. For example, the serial number of the normal card is 0x9C62CAE6 in hexadecimal, or 2,623,720,166 in decimal, which is much more feasible. Very low serial numbers, like very low MAC ID addresses, are hallmarks of a ghost shift.

Finally, the manufacturer's ID on the irregular card is 0x41 (capital *A* in ASCII), which I didn't recognize.* The original equipment manufacturer identification (OEMID) number was 0x3432—an ASCII 42, which is one more than the hex value for the manufacturer ID. Manufacturer IDs are usually the ASCII character given by the hexadecimal value, not the hexadecimal values themselves. Confusing hex and ASCII is a possible sign that someone who didn't appreciate the meaning of the fields was running a ghost shift making these cards.

Were the MicroSD Cards Authentic?

Armed with this evidence, Chumby confronted the Kingston distributor in China and Kingston's US sales representative. We asked whether the cards were authentic and, if so, why the serialization codes were irregular. After some time, Kingston swore the cards were authentic, not fakes, but it did reverse

* JEDEC Publication N. 106AA lists all SD card manufacturer ID codes, and 0x41 wasn't on there.

its position on exchanging the cards. The company took back the programmed cards and gave us new ones, no further questions asked.

However, Kingston never said why the card ID numbers were irregular. I know Chumby was small fry compared to the Nokias of the world, but companies should still answer basic questions about quality control, even for small fry. I was once accidentally shipped an old version of a Quintic part, and once I could prove the issue, I received world-class customer service from Quintic. The company gave me a thorough explanation and immediately paid for a full exchange of the parts. That was exemplary service, and I commend and strongly recommend Quintic for it. Kingston, on the other hand, did not set an example to follow.

I'd normally have disqualified Kingston as a vendor, but I was persistent. It was disconcerting that a high-profile, established brand would stand behind such irregular components. Who could say SanDisk or Samsung wouldn't do the same? Price erosion at the time hit flash vendors hard, and as small fry, I could have been taken advantage of by any of those companies as a sink for marginal material to improve their bottom line. Given the relatively high cost of microSD cards, I needed *incoming quality control (IQC)* guidelines for inspections to follow to accept or reject shipments from memory vendors based on set quality standards. To develop those guidelines, I continued digging for the truth behind those cards.

Further Forensic Investigation

First, I collected a lot of sample microSD cards. I wanted to collect both regular *and* irregular cards in the wild, so I went to the Hua Qiang Bei district and wandered around the gray markets there. I bought 10 memory cards from small vendors, at prices from 30 to 50 RMB ($4.40 to $7.30 USD).

Shopping for irregular cards was interesting. In talking to a couple dozen vendors, I learned that Kingston, as a brand, was weak in China for microSD cards. SanDisk did a lot more marketing, so SanDisk cards were much easier to find on the open market, and the quality of gray-market SanDisk cards was fairly consistent.

Small vendors were also entirely brazen about selling well-crafted fakes. They had bare cards sitting loose in trays in the display case. (Page 11 in Chapter 1 has photos showing what an SD card vendor's stall looks like.) Once I agreed on a price and committed to buying a card, the vendor tossed a loose card into a "real" Kingston retail package, miraculously pulled out a certificate—complete with hologram, serial numbers, and a kingston.com URL to visit to validate the purchase—and slapped the certificate on the back of the retail package right in front of my eyes.

A freshly purchased Kingston microSD card. It was just like new!

One vendor particularly interested me. There was literally a mom, a pop, and one young child sitting in a small stall of the mobile phone market. They were busily slapping dozens of non-Kingston cards into Kingston retail packaging. They had no desire to sell to me, but I was persistent. This card interested me in particular because it also had the broken *D* logo, but no Kingston marking. The preceding photo is the card and the package it came in; the card is Sample 4 in the next section, where you can see a detailed analysis of seven different microSD cards from my shopping trip.

Gathering Data

After collecting my samples, I read out their card ID information by checking their */sys* entries under Linux and then decapsulated (that is, dissolved) their packages with nitric acid. As you can see in the photos in Table 2, my decapsulation technique was pretty crude. Most of the damage to the cards came from removing dissolved encapsulant with acetone and a Q-tip. I had to get a little rough, which didn't do the bond wires any favors. But it was good enough for my purposes.

Here's all the basic information I pulled from those cards:

Sample 1 The irregular card that started this whole investigation. It was purchased through a sanctioned Kingston distributor in China, and to the best of my knowledge, none were shipped to Chumby's end customers. MID = 0x000041, OEMID = 0x3432, serial = 0x960, name = SD2GB.

Sample 2 A normal card that I purchased from the same sanctioned Kingston distributor in China where I bought Sample 1. It was typical of microSD cards actually shipped in the first lot of chumby Ones. MID = 0x000002, OEMID = 0x544D, serial = 0x9C62CAE6, name = SA02G.

Sample 3 A Kingston card purchased through a major US retail chain. MID = 0x000002, OEMID = 0x544D, serial = xA6EDFA97, name = SD02G. Note how the MID and OEMID are identical to those Sample 2, but not Sample 1.

Sample 4 The non-Kingston card I saw slapped into Kingston-marked packaging, bought on the open market in Shenzhen. MID = 0x000012, OEMID = 0x3456, serial = 0x253, name = MS. Note the low serial number.

Sample 5 A device from a more established retailer in the Shenzhen market. I bought it because it had the XXX.A00LF marking, like my original irregular card. MID = 0x000027, OEMID = 0x5048, serial = 0x7CA01E9C, name = SD2GB.

Sample 6 A SanDisk card bought on the open market from a sketchy shop run by a sassy chain-smoking girl who wouldn't stop texting. I actually acquired three total SanDisk cards from different sketchy sources, but all of them checked out with the same CID info, so I opened only one. MID = 0x000003, OEMID = 0x5344, serial = 0x114E933D, name = SU02G.

Sample 7 A Samsung card that I bought from a Samsung wholesale distributor. I didn't scan this one before decapsulating it, and the card actually had no markings on the outside (it was blank, with just a laser mark on the back), so I didn't photograph it. From appearances alone, it was the sketchiest of the bunch, but it was one of the best built. You can't judge a book by its cover! MID = 0x00001B, OEMID = 0x534D, serial = 0xB1FE8A54, name = 00000.

That's a lot of data, and I had my work cut out for me in drawing some kind of useful conclusion from it all.

NOTE *Interestingly, one SanDisk card from three in Sample 6 turned out to be used and only quick-formatted. With help from some recovery software, I found DLLs, WAVs, maps, and VeriSign certificates belonging to Navione's Careland GPS. Someday, I'll acquire lots of refurb microSD cards and collect interesting data from them.*

Table 2: A Breakdown of All the Cards Collected for the Investigation

	Sample 1: Original Kingston card from authorized Kingston distro	**Sample 2: Normal Kingston card from authorized Kingston distro**	**Sample 3: US retail Kingston card**
Front marking			
Back marking			
Decapsulated		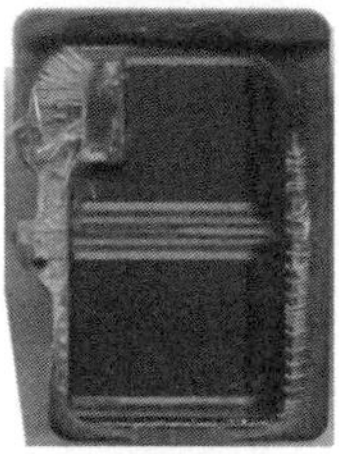	
Controller die marking		 	
Flash die marking		SanDisk/Toshiba flash	

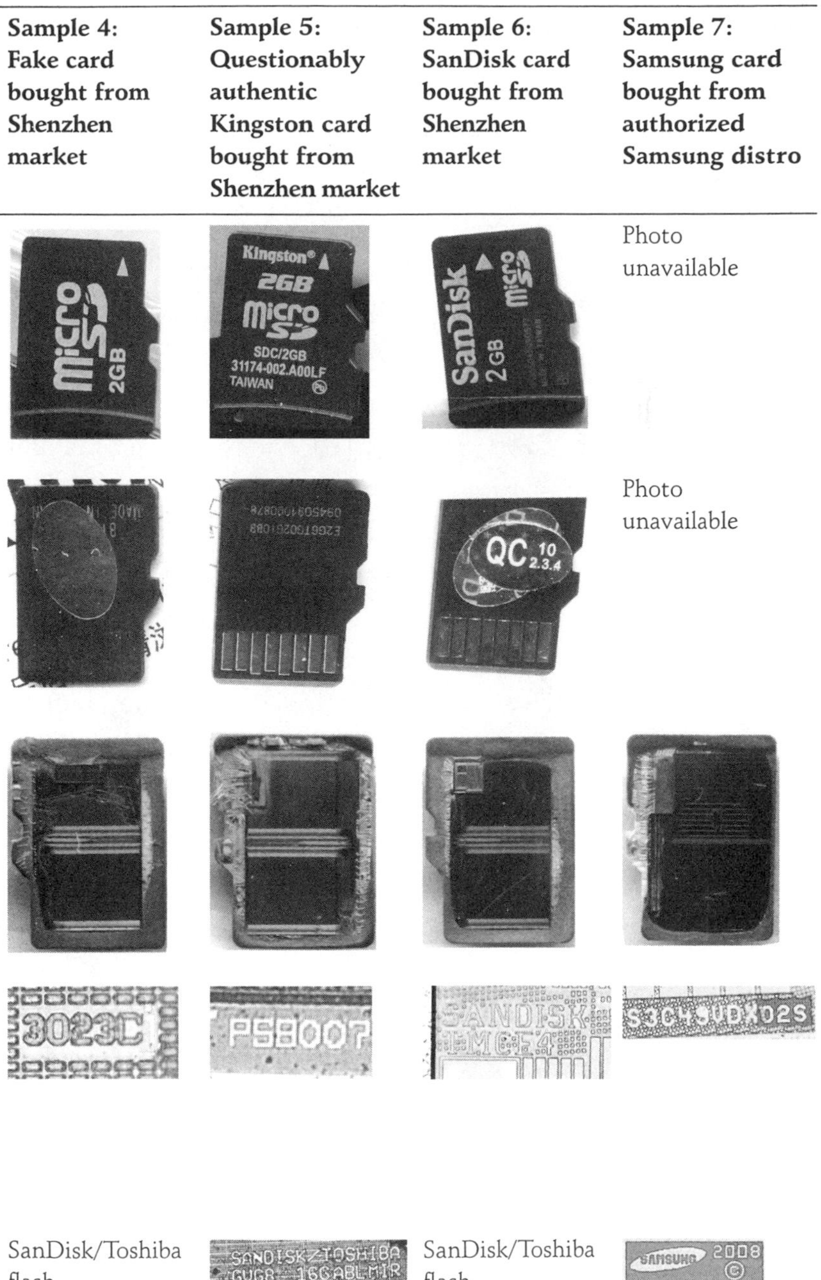

Sample 4: Fake card bought from Shenzhen market	Sample 5: Questionably authentic Kingston card bought from Shenzhen market	Sample 6: SanDisk card bought from Shenzhen market	Sample 7: Samsung card bought from authorized Samsung distro
			Photo unavailable
			Photo unavailable
SanDisk/Toshiba flash		SanDisk/Toshiba flash	

Summarizing My Findings

Here are the most interesting high-level conclusions I drew from my survey:

- The "normal" Kingston cards (Samples 2 and 3) were fabricated by Toshiba, as indicated by the flash die markings and their OEMIDs. In ASCII, 0x544D is *TM*, presumably for *Toshiba Memory*. These cards employ Toshiba controllers and Toshiba memory chips and seem to be of good quality. Thankfully, they were only ones sent to Chumby customers.
- The irregular card (Sample 1) used the same controller chip as the outright fake (Sample 4) I bought in the market. Both the irregular Kingston and the fake Kingston had low serial numbers and wacky ID information. Both of these cards exhibited abnormal operation under certain circumstances. I still hesitate to call Kingston's irregular card a fake, as that's a very strong accusation, but its construction was similar to another card of clearly questionable quality, which leads me to question Kingston's choice of authorized manufacturing partners.
- The irregular card is the only card in the group that does not use a stacked CSP construction. Instead, it uses *side-by-side bonding*—that is, the microcontroller and the memory chip are simply placed next to each other. Stacked CSPs place the microcontroller on top of the memory chip. This is significantly more complex than side-by-side placement because the chips must first have their inert back-side material ground off to make the overall height of the stack fit inside such a slim package. Despite the difficulty, stacking chips is popular because it allows vendors to cram more silicon into the same footprint.
- The only two memory chip foundries in this sample set were Toshiba/SanDisk and Samsung. (SanDisk and Toshiba co-own the factory that makes their memory chips.)

- Samsung's NAND die, which is the most expensive part of a microSD card, is about 17 percent larger than dies from Toshiba/SanDisk. This means that Samsung microSD cards should naturally carry a slightly higher price than Toshiba/SanDisk cards. However, Samsung can offset that against the ability to place the same bare die that normally gets crammed inside a microSD package into thin small outline package (TSOP) devices suitable for board-level machine assembly instead. If demand for microSD cards slumps, Samsung can slap excess bare dies inside TSOP packages and sell those to third parties that do conventional machine assembly of chips. Plus, Samsung also doesn't have a middleman like Kingston to eat away at margins.

I knew (like many others in manufacturing) that Kingston wasn't a semiconductor manufacturer, in that it owned no fabrication facilities, but this research implied that Kingston did no original design of its own. I hoped to at least find a Kingston-branded controller chip inside the Kingston cards, even if the chip was fabricated by a foundry. I also expected to see Kingston sourcing memory chips from a broader variety of companies. Being able to balance the supply chain and be less dependent on a single, large competitor for chips would be a significant value-add to customers, giving Kingston leverage to negotiate a better price that few others can achieve. But every Kingston card I bought had a SanDisk/Toshiba memory chip inside. The only "value-add" that I saw was in the selection of the controller chip.

Oddly enough, of all the vendors, Kingston quoted Chumby with the best lead times and pricing, despite SanDisk and Samsung making all their own silicon and thereby having lower inherent costs. This told me that Kingston must have a very low margin on its microSD cards, which could explain why irregular cards found their way into its supply chain. Kingston is also probably more willing to talk to smaller accounts like Chumby because, as a channel brand, Kingston can't compete against OEMs like SanDisk or Samsung for the biggest contracts from the likes of Nokia and Apple.

So, the irregular microSD card I pulled from the chumby One production line may not have been counterfeit, but it was still a child of the remarking ecosystem in China. Kingston is more of a channel trader and less of a technology provider, and is probably seen by SanDisk and Toshiba as a demand buffer for their production output. I also wouldn't be surprised if SanDisk/Toshiba sold Kingston less-than-perfect parts, keeping the best of the lot for themselves. Thus I'd expect Kingston cards to have slightly more defective sectors, but thanks to the magic of error correction and spare sectors, this fact is hidden to end users.

As a result, Kingston plays an important role in stabilizing microSD card prices and improving fab margins. But the potential conflict of interest seems staggering, and I'm still very curious about how this ecosystem came to be. Buying a significant amount of a competitor's technology from a competitor's fab yet still selling at a competitive price is counterintuitive to me, and perhaps my greatest folly in investigating that irregular microSD card was expecting something different.

FAKE FPGAS

Anyone who has done manufacturing in China for a while will have more than one story about irregularities in the supply chain. Here's another one of my favorite stories, which highlights some of the core incentives that drive agents to cheat.

The White Screen Issue

It was March 2013, and I was wrapping up the first volume production run of a bespoke robotics controller board codenamed

Kovan.* At the conclusion of any production run, I always review the list of issues encountered in production, to identify areas of improvement. Manufacturing is a Sisyphean struggle toward perfection: every run has some units you just have to scrap, and the difference between profit and loss is how well you can manage the scrap rate.

On this run, one particular problem, dubbed the "white screen issue" after its most obvious symptom, was the dominant problem. About 4 percent of the total run exhibited this problem, accounting for almost 80 percent of unit failures. I had the factory send me a few samples of the failed units to analyze in more detail.

As I've often discovered when analyzing failed units, the most obvious symptom of the problem was only tangentially related to the root cause. The LCD screen appeared white on these units because the FPGA failed to configure. An FPGA, short for *Field Programmable Gate Array*, is essentially a blob of logic and memory devices embedded in a dense network of wires that can be configured at runtime to behave a certain way. The behavior of the FPGA is typically described in a high-level language that resembles a programming language like C (for instance, Verilog) or Ada (like VHDL), which is then compiled into a configuration bitstream.

FPGAs are very handy for implementing time-sensitive hardware interfaces that software would have trouble emulating. In this particular application, the FPGA controlled everything from the motors to the sensors and even the LCD. When the FPGA failed to configure, the LCD didn't receive sync and data signals, leading it to show a blank, white screen instead of the expected factory test patterns.

FPGA failure was a big deal. For starters, the FPGA was the most expensive part on the board by a long shot, at around

---

* Kovan is open hardware; you can read more about it and download the source on the Kosagi wiki at *http://www.kosagi.com/w/index.php?title=Kovan_Main_Page*.

$11 per chip. I was also worried this problem could point to a deeper design issue. Perhaps the FPGA's power regulators were unstable, or maybe there was an issue with the boot sequence that aggravated a corner case in configuration timing that would creep into the "good" production units as they aged. The situation definitely warranted a deeper investigation.

Incorrect ID Codes

I hooked up the debug console, dug into the problem, and discovered that the failure was linked to the FPGA not responding with the correct ID code. The ID code is checked via queries over a test access bus known as *JTAG*. Most users don't check an FPGA ID before programming, but we designed an ID code check into Kovan because we allowed customers to specify what capacity FPGA they wanted to use for a given production lot. Some applications are more demanding, while others are more cost-sensitive. As a result, a customer could have a mixed inventory of FPGAs, and we wanted to be able to detect and protect the hardware from an accidental mismatch between the bitstream and the FPGA.

But this was a single production lot, and in theory all the FPGAs should have been the same. How, then, could the FPGA have reported a mismatched ID code at all? I scratched my head for a while and suspected a bug in our JTAG implementation, until I looked up the reported ID code. It was a known code—but for silicon marked as "Engineering Samples" from Xilinx, the vendor that makes these FPGAs. Engineering samples are preproduction units sold by Xilinx that have some minor known bugs but are sufficiently functional for most applications, to the point where most customers wouldn't see a difference, *except* for the ID code.

I looked closer at the PCB, and for the first time, I noticed that a small, white rectangle was laser-etched into the FPGA's surface. The rectangle was right below the part number, where the "ES" designator for an engineering sample would normally be marked. Someone had blasted the letters off and sold us engineering samples as full production units!

An engineering sample FPGA on a Kovan board

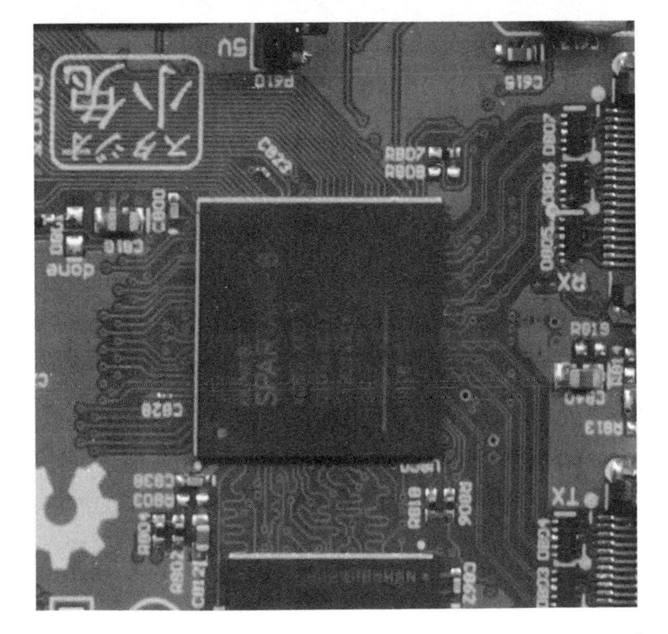

For contrast, an FPGA of the same type that hasn't been tampered with

The problem was very clearly a supply chain issue, not a design issue. Someone in the chain was taking ES silicon, blasting off the letters, and blending them in with legitimate units at a rate of around 3 to 5 percent. Typically, Xilinx would require that all ES silicon in a distributor's inventory be scrapped once production units become available, but the ES units were almost fully functional, to the point where most applications would be unaffected. A production bitstream would seamlessly load into an ES part, and nobody would know the difference. The only way to tell them apart would be by doing an ID code check, which is, as I noted previously, atypical.

Thus, slipping ES silicon into production lots would likely go unnoticed. Mixing ES parts in at a rate of 3 to 5 percent was also very clever: a low mix rate makes substitutions very hard to catch without 100 percent prescreening of the parts. Even in production, if the ES silicon were marginal, it would be maddeningly difficult to nail down the root cause of an issue due to its rarity.

In fact, there's a correlation between manufacturing difficulty and the use of FPGAs. Usually if your design calls for an FPGA, you're pushing boundaries on multiple fronts, so a scrap rate of a few percent is to be expected. The margin on FPGA-powered hardware is also often fat enough that a 4 percent failure rate might simply be accepted by the end customer. Thus, whoever did this knew exactly what they were doing; it was virtually risk-free money.

Finally, it's important to note that most vendors in a supply chain survive on single-digit margins, so finding an extra 3 to 5 percent of "free money" on the most expensive part on a board virtually doubles profitability. That provides a very strong incentive to cheat, especially if you think you won't be caught.

The Solution

The resolution to this problem was quite interesting. I met with the managers and CEO of AQS, the CM charged with producing Kovan, briefed them about the problem, and showed them the evidence I had accumulated. When my presentation ended, the CEO didn't point a finger at upstream vendors or partners. Instead, he immediately looked his staff in the eyes and asked, "Did any of you do this?" He understood better than anyone else in the room that any individual buyer or manager would effectively double their take-home pay that month if they could pull off this cheat without getting caught.

In other words, the truly remarkable part of this situation is how rarely the problem I experienced happens, given what's at stake and how hard these problems are to catch. And while I do have a few good bar stories to tell about fakes in the supply chain, remember that I've also shipped hundreds of thousands of units of good product. The majority of people I've worked with in China are hardworking, honest people who pass on easy opportunities to cheat me and turn a profit. It's important not to generalize the whole based on the bad actions of a few.

At the end of the day, the vendor who sold us the chips didn't admit fault, but they did replace all remarked units at their own cost. (We still had to pay for the labor cost to replace the chips and recertify the boards.) This is about the closest you can get to an amicable resolution in China when you're not a giant like Apple or Foxconn. I did send a note to Xilinx HQ about potential misbehavior by one of their authorized vendors, but in the end, I'm a small customer, and the substitution of parts could have happened literally anywhere on the supply chain. Even the courier delivering the packages could have done the swap.

It wouldn't be worth the cost to Xilinx in terms of manpower, relationships, and focus to investigate the problem and rat out the one bad actor in literally hundreds of possible suspects. But I'd like to imagine that at least a memo was sent around, and whoever was swapping in the ES parts got scared enough that they stopped.

CLOSING THOUGHTS

At the end of the day, a permissive IP ecosystem has benefits and drawbacks. As an engineer and a designer, I prefer to be in an ecosystem where ideas are accessible, even if it means I have to be on guard for occasional problems with fake goods. Put another way, a fundamental prerequisite for virality is the ability to make copies. The explosion of interest in hardware startups is in part thanks to the highly competitive manufacturing ecosystem that could flourish only in a product-over-patent culture.

Westerners who come to China without understanding the principles of *gongkai* and *guanxi** often feel like they're being cheated. But once you understand the rules and learn how to use them to drive your interests, you won't feel like the game is rigged against you anymore.

In the US IP system, honor has little economic value, and law trumps honor. For example, patent trolling is a perfectly legal, and very profitable, way to make a living. In the Chinese system, however, reputation can trump law. This opens the door for corruption but also crowdsources the enforcement of social and moral values, driving a market value for honor, especially in local, tightly knit communities.

Of course, the approach of making money by locking up ideas and selling the rights to them is patently incompatible with a permissive IP ecosystem. Thankfully, the notion that ideas are community property dovetails nicely with my open source philosophies. In the next part of the book, I'll talk more about my experiences creating open hardware and building businesses rooted in these principles.

---

* *Guanxi* (关系) is a traditional social networking platform deeply embedded in the Chinese culture. Like modern social networks, it has notions of followers, likes/dislikes, karma, and moderators. Guanxi predates the modern legal system and can be more effective than the civil code for resolving or avoiding all manners of disputes. Guanxi is also essential in facilitating new deals and relationships.

Part 3

what open hardware means to me

Before there was open hardware, hardware was open.

A yellow, tattered sheet of paper hanging next to my monitors—the schematic for the Apple II computer—reminds me of that fact every day. When I got the schematic as a child, it became a blueprint for the rest of my life. I couldn't understand the schematic, but that didn't matter; it taught me that *hardware is knowable.* It empowered me to understand my world and master the technology I relied on. That empowerment propels me to this day.

The legal doctrine of open source was still nascent when the Apple II was created, so while anyone can read the schematic, it bears no open source license. It simply shows the patent number 4,136,359. Back then, people just shared ideas—until investors with lawyers came along and tragically spoiled the commons. The software community defended itself with the same tools used against it: primarily, copyright law.

Copyright law originally applied to literary and artistic works. Today it also applies to computer code because, like literature and art, code is a form of expressive speech. In the same way that you can copyright a painting of the Grand

Canyon but not the Grand Canyon itself, you can copyright an implementation of Quicksort in C but not Quicksort itself. To ensure source code could be shared freely, the software community created open source licenses. Those licenses range from *copyleft* (that is, openness begets openness) arrangements like the GNU Public License (GPL) to more permissive agreements that boil down to "acknowledge me, don't sue me, and otherwise do as you wish," like the Berkeley Software Distribution (BSD) licenses.

Hardware blueprints can be protected by copyright, too, but blueprints are functional, so defining "open hardware" is trickier. Virtually every piece of hardware used to ship with a schematic. Somewhere along the way, however, it became impossible for users to service hardware themselves without breaking its warranty. Devices are now filled with trade secrets. This shift created an artificial distinction between closed and open hardware. I say "artificial" because while software can be encrypted with ciphers so strong you'd have to build a planet-sized computer to break them, you can reverse any hardware design into a schematic, given a powerful enough microscope and the software to stitch and process the resulting images.

The internet is littered with well-intentioned but misguided attempts to apply software-centric open copyright licenses to hardware. But using a software license on a piece of hardware is like filing a marriage license for a corporate merger: while the license conveys the author's intent, it may not actually do anything. For example, the text of the GPL doesn't use the word *hardware* once, meaning a court could rule that the GPL doesn't legally apply to hardware.

Some hardware-specific open licenses have been created to help rectify the situation (the CERN OHL is a decent copyleft-style hardware license), but the community is divided over how much of the creation process has to be open for a piece of hardware to be considered open. For instance, if I share schematics for a board I designed using a closed-source tool, many would argue that the design does not qualify as open source. But even if I designed the board using a schematic capture and layout tool that was free and open source software (F/OSS) compliant, what about the designs of the silicon chips it uses or the bits of firmware burned into the silicon? Do we need to see blueprints of the particle accelerators used to shoot dopants into the silicon? What about the machine used to engrave the masks used for silicon production? It's turtles all the way down. Hardware can't be purely open source, because at some point, ideas must translate into matter, and access to the objects required to transform and shape matter is rarely open to the community.

There are, however, much more pragmatic approaches to open hardware than doing electron microscopy or demanding open silicon foundries. Simply sharing blueprints at a given layer of abstraction takes much less effort, is more intuitive, and still has a positive effect. The shanzhai's gray-market style of open source, which I referred to in earlier chapters as *gongkai*, reaps the benefits of such sharing. In China, blueprints are shared publicly, but under dubious terms. Most designs still bear "confidential" or "proprietary" copyright notices, and the shanzhai use pirated copies of professional-grade, closed source design software to create derivative works. But at the end of the day, this laissez-faire openness creates an ecosystem where hundreds of small companies make a living repairing or building mobile phones. Walking through the electronics markets of Shenzhen made me realize that building a phone isn't difficult or scary. Communities outside the shanzhai just don't feel empowered to peer inside the box, due to restrictive IP laws.

The gongkai ecosystem, explored in Part 2, values intellectual and physical property almost equally. Schematics without a supply chain are useless: you can't make a phone call with blueprints for a phone. Likewise, chipmakers have no business if no products use their chips. As a result, hardware creators have a natural incentive to share information, particularly the information necessary to design a given module or chip into a larger system. Getting a customer to adopt chip-specific design IP virtually guarantees that customer will purchase the same chips when they're ready to bring a product to mass production. This balance between IP and the supply chain has been difficult to strike in IP-centric Western ecosystems, where ideas are much more valuable than factories. This may partially explain why so many manufacturing jobs have migrated to China, an ecosystem that more comparably values the production of products and the ideas behind them.

I'm optimistic that with consistent effort, growing public awareness, and the right economic conditions, the world's hardware ecosystem will eventually yield an open silicon foundry. However, until then, "open hardware" has to be a more pragmatic concept that is constrained to exist within certain layers of abstraction. After all, just being able to share blueprints (even if the licenses aren't perfect and the formats aren't easily edited) dramatically affects innovation. The shanzhai are living proof.

Whether it's gongkai or open source, open hardware is about empowering users to be the masters of their own technology, not about any specific legal arrangement. Damn the torpedoes—full speed ahead! The freedom to learn, tinker, and improve technology is so core to my person that I view it as a basic human right. Freedom atrophies if not exercised, which is why I actively defend this freedom. I share my work openly, hoping to empower others and raise awareness that technology is knowable. We're not slaves to our computers or the corporations that build them.

I also challenge legislative and legal attempts to curtail our freedoms. I was born into a DMCA-free world; I'd like to leave the world in a similar state by establishing that everyone has the right to understand, repair, and modify the things they own. This is more important than ever as we become increasingly dependent upon technology. If we allow technology to become a black box, we also surrender our agency to the companies and governments that produce and regulate it.

This part of the book describes how I built three open hardware platforms: chumby, Novena, and chibitronics. I hope that by reading my stories, you'll also realize hardware is knowable and be empowered by this knowledge.

6. the story of chumby

One of my earliest open hardware projects was chumby, the Wi-Fi-enabled content delivery device that took me to China to set up my first supply chain in 2007.* Working on chumby was personally exciting to me for two reasons. First, I had the opportunity to build a product that could improve people's lives in some small way. The always-on, always-connected users who blog and rely on IM to keep in touch could use chumby to make those connections more easily. At the same time, chumby was a chance for me to create a truly open platform that enabled hackers to tinker and modify it however they liked.

* Of course, I want to make clear that I wasn't the only guy behind chumby; I worked with a whole team of fun, talented people. As I mentioned in Chapter 1, I was just the lead hardware designer, though I did the Linux kernel stuff too. (That was new for me at the time, but it was a lot of fun learning the insides of Linux from boot to halt!)

A HACKER-FRIENDLY PLATFORM

Hackers have an insatiable desire to extend, modify, customize, and abuse consumer products to discover unintended functionality. At Chumby, we hoped hackers would learn how the device worked and transform it to do things we never imagined, so we designed chumby to be as open as possible to *anybody* who wanted to hack it. We considered not only open source software hackers, but also hardware hackers, artists, and crafters—that is, people skilled with and passionate about noncomputer things, like metalworking, sewing, or carpentry. To encourage and enable chumby hackers, we made the source code, schematics, board layouts, bill of materials, flat patterns, and 3D CAD databases of the plastic pieces freely available. You can still find them all on the chumby wiki (*http://wiki.chumby.com/*).

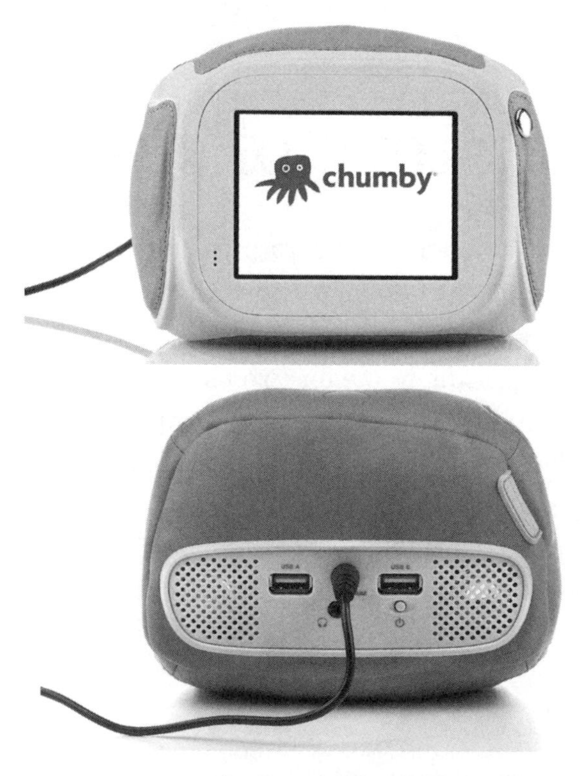

The original soft chumby

The idea was to let hackers break away from point-solution hacks on inscrutable hardware and into hacks they could share with just about anyone. For instance, imagine you add a blood pressure cuff to a chumby and give the chumby to your grandmother. Now you can check on Grandma's health, and she can watch pictures of her grandchildren while she gets her blood pressure taken. But imagine this scenario with a WRT-54G router instead of a chumby. Sure, you can add a blood pressure cuff to a WRT-54G as well (in fact, it's quite similar to chumby architecturally), but try teaching Grandma how to set it up and use it. In other words, we felt making chumby a simple product would allow hackers to make their own hacks more usable and more understandable to the less technical people in their lives.

Making chumby open had other benefits for hackers, too. This time, imagine your thermostat is a little too far from the place where you actually want to regulate temperature. You could solve that problem in a weekend by adding a temperature sensor to a chumby. The chumby platform has Wi-Fi and I built a hacker sensor package for the device, so the project would require minimal hardware grunge work: you'd just mod two chumbys (one with a temperature sensor and one with an interface to the thermostat) and enable both with the sensor package. Such a device would not only help you keep your living room at the right temperature but also tell you the latest news and help you track your favorite TV shows.

The icing on the cake is that you'd also be free to publish your modifications and even resell modified chumbys with those custom capabilities. Others could benefit from your work, and you could make some money. (On a lighter note, the original chumby housing was made of fabric, so you could even modify it to match your décor!)

The original chumby design, now called the *chumby classic*, premiered at FOO Camp in 2006, and it went on sale in 2008.

Unfortunately, however, the chumby classic hit full-stride launch in the middle of the worst economic downturn since the Great Depression. Its cute, cuddly form factor had a price tag that many consumers just couldn't stomach, so I did what any entrepreneur would do in a recession: I scaled back.

EVOLVING CHUMBY

Shortly after Lehman Brothers filed for Chapter 11 bankruptcy protection in 2008, we started work on a product that could address a new economic reality. As I drew my first napkin sketches for the product, which we later dubbed the *chumby One*, the stock market was in free fall and losing several hundred points a day. Given that, the key goal was cost reduction. I took a good, hard look at the whole design so I could build a cheaper, faster product that would be better for the market. We wanted chumby One to win new customers yet retain the loyalty of our existing consumer base, and we wanted it out before Christmas 2009.

Fortunately, an applications engineer from Freescale (since acquired by NXP) contacted me about a new, remarkably inexpensive CPU (the i.MX233) that Freescale planned to launch in 2009. It looked like a promising fit for chumby, so I drew up some straw-man renderings and ran some cost scenarios. At CES in January 2009, we shared the new design with a few potential customers to get feedback on the features and pricing. The idea slow-rolled through March, and after the Chinese New Year, I built the first prototype board.

NOTE *One really cool thing about the i.MX233 is that it has embedded power regulators, and they aren't just linear regulators: they're switching regulators. But they're not just any switching regulators; they derive three voltages using just a single inductor! How cool is that? I have to give mad props to the guy who designed that system.*

Around May, we contracted an industrial designer to do some sketches, and by June, we had a near-final industrial design. We made our first 3D-printed prototypes around then, but we couldn't afford a mechanical engineering contractor. I had to learn SolidWorks and do the mechanical integration for the 3D prototype myself. Since I enjoy learning new things, the experience was quite rewarding.

In July, we inked a purchase order for steel tooling, and by August, we had first-shot plastics. I spent September refining and debugging the design and October on more testing, refining, and ramping up mass production. By November 2009, the first shipment of chumby Ones was 35,000 feet above the Pacific Ocean en route to LAX.

The finished chumby One

The chumby One retailed for about half the price of the chumby classic, and it had more features, like an FM radio and support for a rechargeable lithium ion battery, a feature users of the squishy, leather chumby classic often requested. The initial reactions to the battery in the chumby One were an interesting study in consumer psychology. For some reason, even though the chumby One was smaller and lighter than the chumby classic and did exactly the same things, people didn't feel it should have a rechargeable battery. They had no intrinsic desire to pick up the chumby One and carry it around. That just goes to show how much form factor influences a consumer's perception of function!

At any rate, customers certainly liked all those options, but to me, they weren't the most significant new features.

A More Hackable Device

What really excited me about the chumby One was that it was much more hackable than the chumby classic. On the chumby classic, we used a soldered-down SLC NAND chip, which was cost-effective but made development quite complicated. Developers were exposed directly to all the warts of NAND flash memory, including bad blocks and error correction, and if the system failed to boot correctly, one had few recovery options. We addressed these problems on the chumby One by storing the firmware on a microSD card.

If you happen to get your hands on a chumby One, you'll notice that you can't replace the microSD card from the outside. We made that choice to prevent nonhackers from pulling the microSD card out and wondering why the device wouldn't boot. But if you unscrew and remove the back panel (no glue seals, unlike the chumby classic), the microSD card is easy to access. Thanks to this key change, hackers didn't have to worry about bricking their chumbys. If someone screwed up the firmware, they could just pull the microSD card out, mount it on their dev box, and write a new image.

We also chose to make the chumby One's microSD card a *managed* NAND device so that we could directly drop ext3 (a popular default Linux filesystem configuration) onto it. The root partition was still mounted as read-only at the factory to prevent accidental damage, but a managed NAND system made remounting the root partition as read/write and modifying the Linux system trivial. We consciously made the OS image use only a small portion of the total microSD card capacity, leaving hackers with over a gigabyte of extra space to load custom applications and libraries. (Keep in mind that a gig was a big deal at the time.)

In hardware, what's good for hackers is also good for developers. The flexibility we added for hackers allowed us to add a ton of great features to the OS. For example, the chumby One supported certain 3G modems and could serve Wi-Fi as an access point through those 3G modems. That basically made the device a 3G-to-Wi-Fi router, which I found enormously useful when I was traveling and needed to create a Wi-Fi hotspot for other devices. We didn't expose that feature at the mainstream user level at first, but we knew we (or anyone else—it was an open project, after all) could wrap a GUI around it and make it more user-friendly if people liked it. And if you plugged a USB keyboard into a chumby One, it would automatically open a console shell that you could type into. That's handy for times when you can't SSH in, like when you're debugging network scripts.

Hardware with No Secrets

As with the chumby classic, we also made the chumby One design as open as possible. We posted schematics, gerber files, and the GPL source code online. In the following figure, you can see a preproduction pilot chumby One board. The mass-production board was basically identical, with some minor tweaks to enhance compatibility with the SMT machines we used in China.

In particular, notice the pair of test points on the board labeled *SETEC ASTRONOMY* in the bottom-left corner of the photo of the back of the mainboard. You could use those points to bypass the write protection on the chumby One's authentication ROM and wipe out the keys that Chumby used to authenticate the device. I can't think of a real reason to do that, but I added them on the principle that hardware you own shouldn't hold secrets from you. If you don't like having encrypted access codes on a device, you should be able to nuke them. In the case of a chumby One, that meant you'd no longer have the codes to fetch widgets from Chumby's servers, but hey, it's your hardware. When hardware is truly yours, you can void the warranty and do what you want with it. Of course, we published the security protocol that chumby Ones used to fetch widgets, too.

I also designed the chumby One motherboard with mounting holes and features so it could be retrofitted back into a chumby classic. Although Chumby never planned to put chumby One boards into chumby classic enclosures—hand-stitched Italian leather was just too expensive, and there were a couple of technical issues with integration—I thought intrepid hackers would appreciate the option to do it themselves.

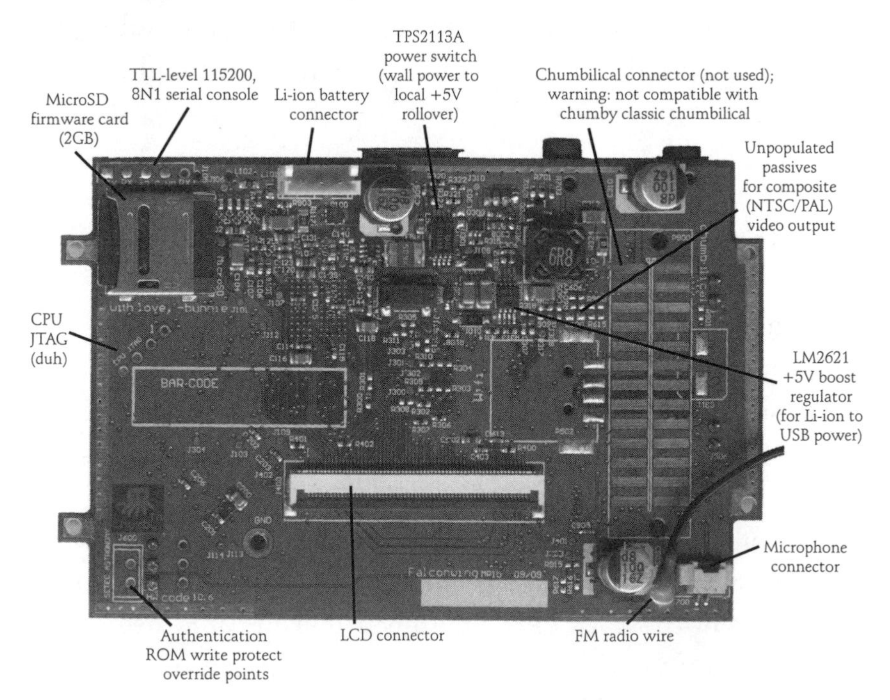

The chumby One mainboard (back)

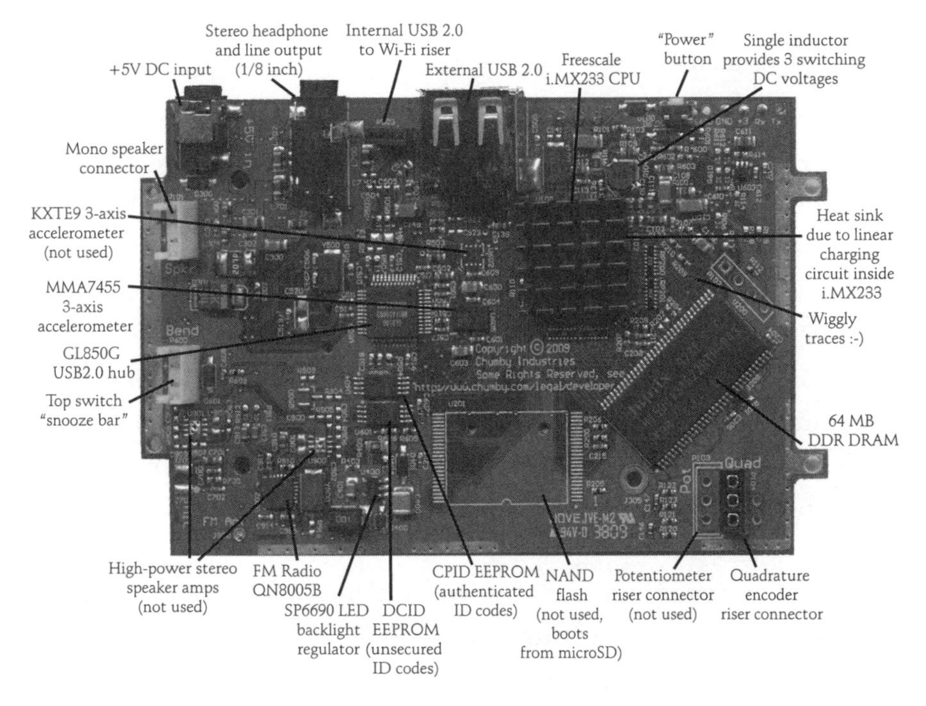

The chumby One mainboard (front)

I continued to work on improving the chumby line for several years, but eventually, I wanted more time for personal projects and a break from entrepreneurship.

THE END OF CHUMBY, NEW ADVENTURES

In April 2012, Chumby as the world knew it came to an end. We had run out of money, and the investors had run out of patience. I'd already left the company discreetly in January; I had a good run, but it was also time for me to move on. Upon hearing the news, my good friend Phil Torrone from *Make:* reached out to me for an interview, and I was happy to oblige. You can read the full interview online,* but I've excerpted parts of it here that you might find useful if you're excited to get into the hardware business.

* See *http://makezine.com/2012/04/30/makes-exclusive-interview-with-andrew-bunnie-huang-the-end-of-chumby-new-adventures/* for the full interview.

Phil: How did you get involved at Chumby? And what was your role at the company?

bunnie: I was originally an advisor to the company, a consultant brought in to figure out some bits of the hardware strategy. We had weekly dinners where we'd talk about what the product might be. Eventually, I got excited enough about the product that I just hammered out an initial prototype motherboard in my spare time. Around the same time, my boss at my prior company was really irritating me (he lectured me about the importance of being in my chair every morning by 9AM, completely ignoring the fact that I'd worked until midnight the day before), so I resigned on the spot and joined the founding team of Chumby.

My role at the company was initially VP of Hardware, which sounds grand. But when the hardware organization consists of exactly one person, you're also the solder jockey and the janitor. Now that I think back on it, the team took a big chance on me. At the time I had no experience in supply chain management and had never been to China. They took a leap of faith and gave me the opportunity to figure it all out. I really appreciate that they gave me so much latitude to learn on the job.

Phil: What was the best part of making the chumby?

bunnie: There were so many great things about making the chumby. I think overall, one of the best parts was that I had to figure everything out from conception to distribution. It meant that I got to see every part of the process firsthand: industrial design, electronics design, tooling, supply chain, retail, and reverse logistics. There are so many things that go into a product, and satisfying that curiosity about how things are made was great.

The other thing I really treasure from making the chumby was all the wonderful people I got to work with and meet along the way. I made a lot of friends, and I had so many excellent mentors.

And finally, I think the best part about making chumby isn't really the making. It's seeing people use it, and seeing people enjoy and appreciate the device. The smile on a user's face is the ultimate reward.

Phil: Can you talk about making a device from start to finish, from idea to factory to retail shelves?

bunnie: One of the best parts about making a device from start to finish is that you have a totally unconstrained set of tools to solve the problems at hand. You can solve business problems with board layout, and vice versa. For example, there was a question about how we could uniquely and flexibly brand units, in a fashion that allowed for swappable faceplates (that is, snap on the NFL faceplate and get your football scores, snap on the Bloomberg faceplate and get your financial news, and so on). This is a topic that could take dozens of meetings to hash out. But as the sole hardware guy, I knew that embedding an EEPROM costs only $0.20 and while everyone else discussed possible solutions in the staff meeting, I fired up my board design tool, added the eight-pin EEPROM to the board, tossed on an appropriate connector, and had the whole solution engineered by the time action items were assigned. It actually took me longer to convince them that the work was done than it took to do the work.

I think I ended up absorbing many of the skills required to build a product from start to finish because it's very difficult to communicate requirements. The question was always whether it would be faster for me to do it myself or to explain it to someone else, wait for them to do it, and possibly have to re-explain it and have them change it. That's one reason I learned mechanical design; the industrial design and plastics tooling is a long pole in the tent for many consumer products, and being able to efficiently and effectively communicate with a mechanical engineering team using their language was important to getting the job done right.

Phil: What were the challenges with retail sales?

bunnie: Retail and distribution were the most difficult challenges. Here are a few difficulties I encountered:

Dealing with the merchant buyers. Brick-and-mortar retailers hire teams of buyers assigned to monetize shelf space. They think about products in terms of revenue per shelf space, and they don't really see anything beyond that. This puts into sharp relief any improvements you want to add to the product that also drive up product costs. Merchants tend to look at your product as so many grams of plastic and so many wires. They multiply those numbers by the commodity price of the raw materials to set expectations for how much they'll pay to have it on the shelf. It's possible to cut better deals, but educating a merchant about the value of your product takes a lot of effort. Unfortunately, the turnover in merchant staff can be fairly high, so you may spend months cutting a deal only to find that the person you were working with has left the organization.

Margin. Everyone in the supply chain has a hand out: the distributor, the merchant, and the factory. Beyond that, market development funds and other slush money have to be factored in. At the end of the day, the shelf cost of a product is about three times your BOM cost. This means adding a $0.50 part turns into a $1.50 retail price impact.

This is aggravated by the fact that prices are quantized into "magic" numbers (like $19.99, $49.99, or $99.99) that you have to hit. You just don't MSRP a product for $127.45. If a product retails for above $99, it's psychologically binned with the $149 or $199 products. When your product's BOM cost approaches one of these quantization points, you'll do lots of soul searching about whether it's worth $0.50 to improve, say, the speakers. Either that small cost increase will come out of your own margin, or you risk pushing your product into a higher price tier.

Cash flow. Retailers are notoriously bad at paying on time. You may negotiate 60-day terms, but often you're not paid after 90 or even 120 days. If your product doesn't sell out so that the retailer has to place another order with you (at which point you have some leverage to collect outstanding payment), you'll get strung out. This can be partially mitigated with financial instruments such as factoring insurance. Insurance companies will sell insurance on anything, including insurance hedging against retailers not paying on time or going insolvent before they can pay you.

Reverse logistics and returns. Many retailers offer no-questions-asked return guarantees. That's great for the customer, but guess who services those returns? The retailer passes the buck back to the entrepreneur! This is part of why payment times can be quite bad: retailers are retaining cash to hand back to customers to satisfy returns. Once the returns are processed, you get to figure out how to get the returned material off their dock and back into a facility where you can refurbish the units. Typically, most returned units aren't defective. They simply didn't meet customer expectations, or the customer had buyer's remorse after an impulse buy. The otherwise working units are usually missing accessories or are cosmetically marred, thereby requiring extensive rework to refurbish.

Contracts. Retailers will hand you a default contract full of terms that very strongly favor them in almost every contingency. Sometimes, the contracts can expose you to liabilities that you can't possibly hope to cover. For example,

I've seen language such that if an affiliated content website was down for longer than a specified amount of time, then you could be liable for nonspecific damage to the brand reputation of the retailer selling your goods. Those sorts of open-ended liabilities are unacceptable, and negotiating them out can take months. Other onerous terms include penalties for late shipments or fines for defective units. The contract negotiation process is very distracting to top management and can put a real drag on an organization.

Phil: Did you get any patents? How do they work within the world of open source?

bunnie: Yes, I actually was granted several patents during my tenure at Chumby. Patents are a very natural way to protect hardware ideas. As F/OSS [free and open source software] licenses like the GPL [GNU general public license] and BSD [Berkeley software distribution] rely on copyright for power, open hardware licenses can likewise draw upon patents for power.

When we started, no license existed that addressed the patent issue, so chumby created its own flavor of open source license. It was basically an automatic cross-license with users who created derivative works. Those who utilized our source would get a license to the patents, under the condition that any patents granted for the derivative work also had to be automatically licensed back to us.

The license had a couple of other restrictions that were not "truly" open, like a condition that the derivative work had to at least give users the option to run the chumby network in a competing product (an opt-in checkpoint during the boot process). There was also an "ask us if you want to manufacture" clause, which stated that derivatives going to mass production had to get additional authorization from Chumby. We added that primarily to create a checkpoint to verify interoperability with the servers, and also to enforce proper trademark and branding rules. Burying that clause in the license meant that the license couldn't be called open source because Chumby could always say no, though it never did in practice. However, the situation does highlight an ongoing struggle in open source hardware: how to address trademark and interoperability issues in an increasingly complex and diverse ecosystem.

Also, the rights to the patents I created at Chumby are all assigned to the investors. They will likely be sold to the highest bidder, which could very well be a patent troll. I would regard that outcome as unfortunate, but it's a reality that I must accept. The investors have the right to explore all lawful venues to recover their investment. In an ideal world, however, I'd buy back the rights at an affordable price, license them to the open source community, and try to establish a material precedent on how to handle patents in the open source community.

Phil: Do you have any advice for a maker who is considering taking venture capitalist funding? Anything different if they're doing open source hardware?

bunnie: I think VC funding is suitable only for accelerating certain kinds of growth. It's not very good for early-stage research and development or businesses that have slow, but steady, growth models.

The hardware model is radically different from the software model. Software is innately scalable. You can acquire 100,000 users overnight. Monetizing the user base in software is trickier, but most software plays start with scale and then worry about money.

Because hardware requires the movement of atoms to acquire a user, scalability is limited by the rate at which you can economically and reliably assemble your atoms and ship them to the customer. On the other hand, there is a very natural point for monetization in hardware: the margin you charge on every unit sold. Money comes earlier and more often, but the growth rate is limited by pesky things like the laws of physics and the availability of raw materials and skilled labor to build the units. Notable exceptions to this rule are concepts like the Square reader. Square's hardware was cleverly designed to be so cheap that its cost was arguably lower than the cost to acquire a customer through other means (like print advertising and mailing campaigns), making the dongle cheap enough to just give away.

Therefore, in hardware, first ask this: what is your distribution channel, and how hard is getting your product to end users? Ultimately, the size of that pipe and the monetary drag on transactions limits the growth rate of your idea. You also have to factor in *boomerang* costs like returns and customer support costs. You'll be shocked at how many support calls you get from people who forgot to plug your product in.

If you have an awesome distribution channel, a solid marketing campaign, and customers lined up out the door, maybe VC is a reasonable match. But a typical maker will start out selling stuff online, possibly in boutique stores. The time it takes to turn capital into revenue will be on the order of months initially, and that's a brutal cycle to finance with VC. All the money you have tied up in the supply chain isn't adding any value to you, but you traded a lot of your ownership in the company to get that money.

I would typically recommend that a maker try to first fund research and development out of pocket, or with a very friendly angel loan. Once you have a prototype and a solid plan for production, it's smarter to go into debt to finance small batches of builds so you're never overextended and build your market one step at a time. Every time you turn inventory, you should come back with more cash, which you can plow into making more inventory.

Doing this forces good discipline. It will help you focus on leaning up the supply chain so that inventory turns faster. The best hardware companies turn inventory in a matter of days. If you're growing your capital base by 20 percent with every inventory turn, it only takes four turns to double your money: $100 turns into $120, which turns into $144, which turns into $172, which on the fourth turn results in $207. That's the magic of compounded percentages.

If you can do a full turn of inventory once every eight weeks and sustain a 20 percent growth rate with each turn, you'll grow your business by over 300 percent in one year. Of course, the markets are never so ideal and predictable, but you can play with turn time versus margin available to grow your business. Higher-margin businesses can take longer to turn inventories and still sustain a palatable growth rate.

Bootstrapping like this is a lot of hard work, but at the end of the day, you own every penny you make, as you have no investors. The glory stories for this model aren't as big as, say, Instagram or Google, but if you're doing it right, you're in control, and your work is more likely to pay off in the end. In fact, many successful Chinese hardware manufacturing businesses grew primarily using bootstrapped funding just like this.

Phil: What are your thoughts on Kickstarter for funding?

bunnie: I don't think it's a good idea to fund early research and development with Kickstarter or other crowdfunding platforms because of the hard commitments you have to make to customers early on. Kickstarter is a great phenomenon, but you also need to be careful raising money there. To some extent, Kickstarter is the ultimate dumb money. Customers are sold on a vision and buy in early on, and you have to deliver on that vision. In crowdsourcing your money, you've also crowdsourced your board of directors. But the road to product development is never smooth. As a result, Kickstarter money can lock you into commitments early on that you can't back out of.

I think Kickstarter can be a better solution than VC, but you should only use it after the idea has matured sufficiently and you're primarily looking to find a better way to finance production than VC money or a bank loan. In fact, after you consider the frictional losses of extracting money from Kickstarter, a bank loan with a few percent interest could be favorable. But of course, a bank loan doesn't come with the same visibility, marketing, and upside potential as a crowdfunding platform.

Phil: When you advise companies, what do you most often suggest to the founders?

bunnie: Ship or die! Particularly if you've accepted VC funding. The moment VC money hits your books, you're on a fixed-length fuse. If that fuse runs out and you haven't created substantial value, a bomb goes off that wipes out a chunk of your valuation. If you've raised a million dollars and you plan to burn it in a year, every day "costs" you $4,000. I use that as a value barometer to guide decision making: if $30 in expedite fees can pull in the schedule on a long-pole task by one day, the money is well spent. This is also part of the reason I lived on "China time" while chumby was in production even though I was in California. Staying up until 4 or 5AM every night to flip emails with the factory and shorten the longest pole in the tent shaved days off the schedule, which translated to tens of thousands of dollars in burn.

In the face of "ship or die," don't look to ship the perfect product. Shipping a product that's good enough is more important than shipping a great product late, especially in consumer electronics or any similarly seasonal business. In consumer electronics, up to 90 percent of your business can happen in the fourth quarter. If you miss Christmas, you'll have no revenue for the next three quarters; missing Christmas is like dropping an extra year of burn on your capitalization table. Worse yet, during that year, your competitors will continue to improve.

Chumby suffered from precisely this. We premiered an alpha version of the device in August 2006, but we missed Christmas 2007. We didn't launch our squishy, connected alarm clock until just after Christmas, in February 2008.

Consider some world events that happened around these dates: the iPhone shipped in June 2007, and the global economy crashed in October 2008. It was bad enough that we had to weather almost a full year, from February 2008 until Christmas 2008, burning venture money to stay warm. But when the economy fell out, so did the appetite for a $200 stocking stuffer. We had too much inventory and had to fight for survival.

If my memory is correct, we could have shipped a product for Christmas 2007. It just wouldn't have been quite as polished and would have lacked some features. But maybe it would have been good enough. In retrospect, the iPhone had by far less momentum in 2007 than in 2008, and we probably could have cleared a lot of inventory. On the other hand, perhaps knowing the iPhone, its apps, and its awesome touchscreen would obsolete a connected alarm clock drove us to second-guess our strategy and delay launch to strengthen features like streaming music integration.

At any rate, the lesson is clear enough to me: ship or die!

A second piece of advice I'd give to hardware companies is to aim high with price. It's virtually impossible to raise your pricing if you start too low, and there's nothing like a sale to get people to buy.

Hardware startups that principally sell online are tempted to set the price as low as possible to drive buzz and improve initial sales. The temptation to sell your $35 device for $49 direct online is huge. After all, that's about a 28 percent margin (unless your BOM doesn't factor in soft costs). That's great, until you've dropped off the front page of Engadget and your sales are plummeting.

Engaging a retailer may help bring in more, and more consistent, sales, but a retailer will initially try to buy your product from you for between 40 and 60 percent of your MSRP. This means they'd want to buy a product for $49 and sell it at $99. If you've already sold a bunch of units at $49, there's no way the retailer can sell it for $99. To access retail, you'd have to sell your $35 product to a retailer for $25 so the retailer can sell it at your established price of $49. Even if you're successful with such a drastic cost-down, you're still left making no money!

Selling your $35 device for $99 might garner fewer customers at first, but your initial margins would be spectacular, and you'd have the room to cut in a retailer or run sales of your own to get more customers. That's part of the reason MSRPs are so high. Retailers also love to use sales to make units move, and a $99 unit priced down to $69 feels like a smart buy. But at $69, the retailer is only making 29 percent margin.

Aiming too low on pricing effectively robs you of the opportunity to use retail as a possible distribution channel, and you simultaneously lose the opportunity to have sales and promotions yourself. Promotions are important because viral marketing can only get you in front of a customer once or twice at best. So when you put your heart and soul into your product, price it like you mean it.

Phil: If you could do it over, how would you change the hardware of the chumby? The software? The way chumby was made?

bunnie: Well, as my previous answer indicates, I would have focused much more on shipping on time, perhaps at the expense of jettisoning some features.

A more counterintuitive thing I learned is that accessories and packaging can take more time to develop than a product. The squishy chumby classic came with a wonderful set of linen and microfiber bags and rubber charms. (We developed over a dozen charms in all.) There was also a custom power adapter, branded ribbons, gift boxes, branded tissue paper . . . I even had to iterate the hardware design and spin an injection-mold tool to improve the attachment method for the charms to the device. I spent at least four months intensely focused on the accessories and packaging for the product. Our fan base went wild over the attention to detail, and that helped goose sales.

But in retrospect, I wonder if we could have done better forgoing the details and shipping before Christmas. One of the most gut-wrenching realizations that small companies have to make is that they aren't Apple. Apple spends over a billion dollars a year on tooling. An injection-molding tool may cost around $40,000 and take two to three months to make; Apple is known to build five or six simultaneously and then scrap all but one so they can evaluate multiple design approaches. For Apple, tossing $200,000 in tooling to save two months' time to market is peanuts. But for a startup that raised a million bucks, that's unthinkable. Apple also has hundreds of staff; a startup has just a few members to do everything. The precision and refinement of Apple's products come at an enormous cost that is out of reach for startups.

I don't mean to say that design isn't important. It's still an absolutely critical element to a product, and good design and attention to detail allow a startup to charge more for a product and differentiate themselves from competitors. Apple has raised the bar very high for design and user experience, and users will judge your product accordingly. But it's important to keep in mind that your true bar for comparison is other startups, not Apple. If your chief competitor is Apple, either you need a billion dollars in cash to invest in product design or you need to rethink your strategy.

That leads to another thing I'd probably change. Pivoting is so important for a startup. A startup has to be able to run circles around big companies. Culturally, Chumby just found it challenging to be agile enough to adapt to a rapidly changing technological landscape.

Of course, hindsight is 20/20. There's a lot we could have done differently, but when I think back on all the early decisions we made and how we got there (the resistive touchscreen, lack of integrated battery, using Flash as our core platform), I don't see how we could have made any different fact-based decisions back then.

But that does show a flaw of fact-based reasoning. Engineers love to make decisions based upon available data and high-confidence models of the future. But I think the real visionaries either don't know enough, or have the sheer conviction and courage to see past the facts and cast a long shot. It's probably a bit of both. Taking risks also means there's a bit of luck involved.

I certainly have a fact-induced myopia. My recent focus on operational efficiency, schedules, and risk management has sapped my ability to have creative and audacious visions. I'm actually taking a year off from entrepreneurship to decompress a bit and to try to rediscover and develop the creative bits of myself that have atrophied over the past couple of years.

Phil: Now that you've been part of a full cycle of a VC-funded company that makes hardware, what suggestions do you have for company structure, from the people to the location to the overall organization?

bunnie: The structure really depends on the type of product you're trying to build. Hardware has many different specialties (like consumer, medical, and industrial) and markets (like high-end boutique, hobby items, and mass market devices). There's good business potential in all of them, but your location, focus, and team composition need to be tuned based on your product and what gives you a competitive edge. At Chumby, hardware was just a barrier to entry for apps to run in your home, so it was instantly a race to the bottom. The hardware part of the company had to run lean (remember, Chumby had one hardware engineer and one operations director), and it needed a China-centric strategy from day one.

Generally, if you can suffer doing a hardware startup through bootstrapping, it's worthwhile. A broad range of hardware products can be bootstrapped at first—and then Kickstarted, debt-financed, or VC-funded to scale. For instance, MakerBot developed and shipped its 3D printer entirely on angel money, before closing a round of VC funding. Bre Pettis, one of the cofounders, once mentioned that they lived on nothing but cup ramen noodles for a month.

Any hardware company that has passed the idea phase and is entering the scaling-up phase has to be razor-focused on operations and cash flow. Maintaining a build-to-order paradigm is critical but difficult: a key metric for any hardware company, small or large, is how quickly you can turn inventory into cash. There are two halves to the equation. One is leaning up your supply chain and trimming lead times so you don't need to sit on much inventory, yet can satisfy new orders quickly. The other is leaning up your cash management so you can bill customers quickly while stretching your credit lines as far as possible. That's a multidimensional optimization problem that can make your head explode without the right staff, so your team should include a crack operations director and someone adept in semi-exotic financial instruments like factoring insurance, collateralized lines of credit, and trade contracts.

Being able to access China effectively early offers a disruptive advantage to your startup (it's hard to ignore the order-of-magnitude advantage China has over the United States in assembly costs), but working with China does come at a huge cost and risk to the organization. It may not be for everyone, particularly on day one.

I outsourced myself to Singapore to get closer to China, because I knew I'd never be able to get away from the China ecosystem. China has such a firm grip on hardware manufacturing, and I think it will take decades for them to lose their edge. This geographic diversity also means that any effective hardware startup has to be able to function effectively with a delocalized team.

Phil: What's next for bunnie? What are you most excited to do next?

bunnie: That is *the* question for me! I don't really know what's next. As I noted earlier in the interview, I'm taking a year off to do things that aren't specifically entrepreneurial. My current priorities are to first have fun with my work, second to not lose too much money, and third to do something good for the community through a combination of hacktivism, volunteer work, and open source methodology. I'm hoping in this year I'll collect the bits of my soul that I've lost along the way, find some new ones, and relearn the value of magic in my life. I'm also spending a fair bit of my focus tuning up myself, getting fit, changing my diet habits, and losing weight. The coolest piece of hardware you'll ever own is your body, and if that's not working well, there's no hope for anything else. Once I'm done with my aimless wanderings, hopefully I'll have a better idea of what's next!

While reviewing that interview for this book, I chuckled a bit to myself. By that point, the year I took off had turned into four years. Several concerned associates of mine asked, "When are you going to stop your midlife crisis and get a real career?" But in retrospect, not going back to the corporate world was the best decision I ever made.

I do live a lot leaner than I did when I had VC/corporate backing, but I have a lot more independence. It was a choice between golden handcuffs and an Aeron chair, or a rucksack and an interesting spot near the horizon. I'm still working on collecting the bits of my soul, and I'm still slowly relearning the values of enchantment and wonder. But at least I have the freedom to contemplate values other than the wealth of my invested shareholders. Thankfully, I had some success in revising my dietary habits and fitness level; tuning up my own body was an excruciating year of calorie tracking, sore muscles, and blistered hands, but it paid off in spades. My mother used to tell me that without health, you have nothing; she's absolutely right. If you don't have the stamina to work, it's hard to turn opportunities into outcomes. With any luck, my health will hold out, and I'll have many more stories to share with you in the future.

WHY THE BEST DAYS OF OPEN HARDWARE ARE YET TO COME

One of the most critical outcomes from my year of soul searching was the realization that the best days of open hardware are still ahead. As I contemplated in my interview with Phil, Chumby didn't fail because of its open hardware model. At worst, the model had little bearing upon the consumer appeal of the product; at best, it was a good talking point. Nowhere in that interview did I gripe about plummeting sales in response to cheap clones appearing on the market due to our liberal open source policies.

Rather, one of our biggest challenges was an inability to keep up with Moore's law. Chumby simply didn't have the resources as a startup to keep pace. It took two to three years to push a major platform revision, at which point that revision was already obsolete. My PhD dissertation* was centered on Moore's law and its impact on computer architecture. The most powerful computers are descendants of a processor designed in the 1970s (the Intel 8085) with derivatives still used today

---

* *http://bunniestudios.com/bunnie/phdthesis.pdf*

as the brains of toaster oven. Why? Because running existing code on backward-compatible CPUs has almost always been faster than porting old code to a new microarchitecture. Given that fact, in my thesis, I designed a microarchitecture that nobody could possibly implement at the time but that might be optimal for a computer that could be built 10 to 15 years out. A small team of researchers would have ample time to develop the infrastructure necessary for a novel computer that would be relevant the day it's finally switched on. I spent several months in the late '90s studying the underpinnings of Moore's law, trying to understand where it runs thin and where it holds strong. At the time, the strongest limitation was the speed of light, so my thesis revolved around architectural tricks to reduce communication latencies.

In 2011, about a decade after my graduation and right around the end of Chumby, I had an opportunity to give a "vision" keynote at the Open Hardware Summit. I decided to review my notes from college and see if there might be another decade left in Moore's law. There isn't, and that has profound ramifications on the future of open source hardware. This section is an adaptation of a blog post I wrote in 2011 sharing my thoughts; thankfully, here in 2016, I've yet to retract any of the statements I made back then.

Where We Came From: Open to Closed

Open hardware is a niche industry, and certain trends have caused the hardware industry to favor large, closed businesses at the expense of small or individual innovators. Looking 20 to 30 years into the future, however, I see a fundamental shift in trends that can tilt the balance of power to favor innovation over scale.

As I said in this part's preface: in the beginning, hardware was open. Early consumer electronic products, such as vacuum-tube radios, often shipped with user manuals containing full schematics, a list of replacement parts, and instructions for service. In the '80s, computers often shipped with schematics. For example, the Apple II shipped with a reference manual that included a full schematic of the mainboard, an artifact I credit for strongly influencing me to get into hardware.

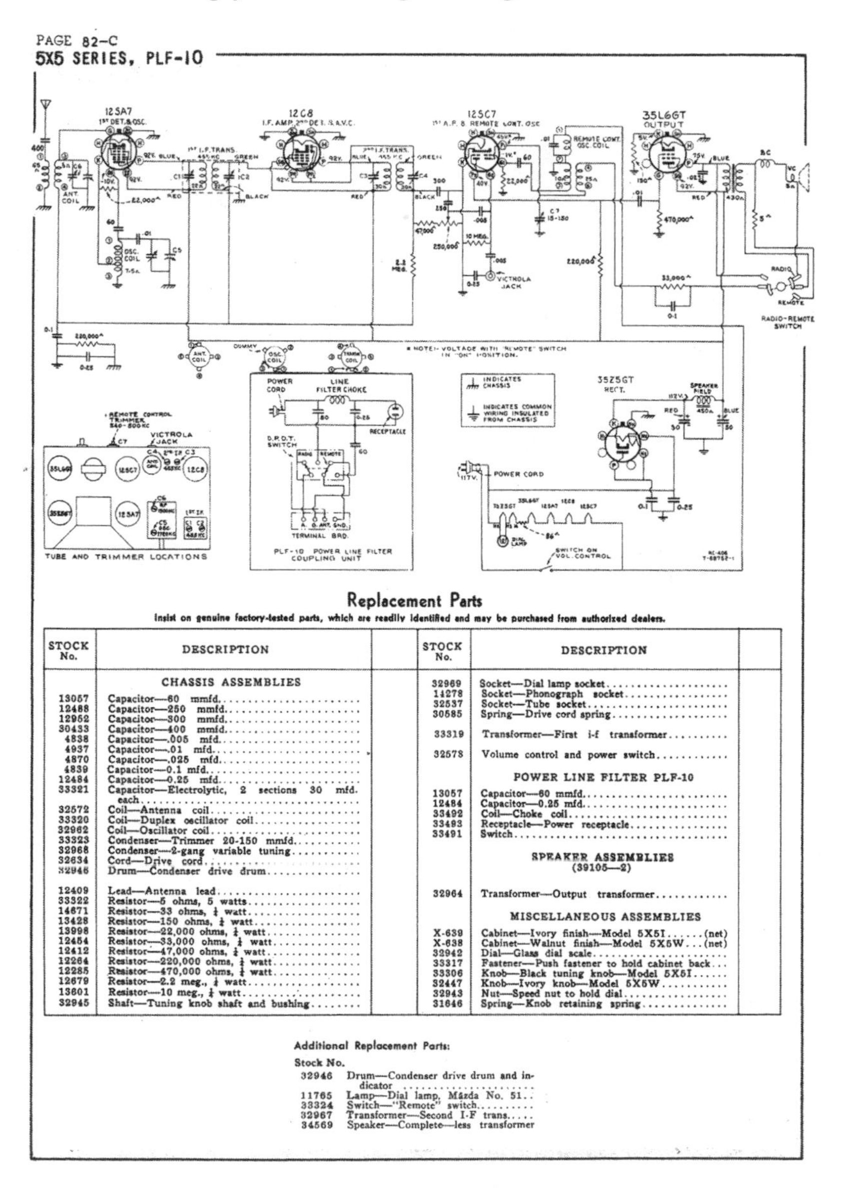

PAGE 82-C
5X5 SERIES, PLF-10

Replacement Parts

Insist on genuine factory-tested parts, which are readily identified and may be purchased from authorized dealers.

STOCK No.	DESCRIPTION
	CHASSIS ASSEMBLIES
13057	Capacitor—60 mmfd.
12488	Capacitor—250 mmfd.
12952	Capacitor—300 mmfd.
30433	Capacitor—400 mmfd.
4838	Capacitor—.005 mfd.
4937	Capacitor—.01 mfd.
4870	Capacitor—.025 mfd.
4839	Capacitor—0.1 mfd.
12484	Capacitor—0.25 mfd.
33321	Capacitor—Electrolytic, 2 sections 30 mfd. each
32572	Coil—Antenna coil
33320	Coil—Duplex oscillator coil
32962	Coil—Oscillator coil
33323	Condenser—Trimmer 20-150 mmfd.
32968	Condenser—2-gang variable tuning
32634	Cord—Drive cord
32946	Drum—Condenser drive drum
12409	Lead—Antenna lead
33322	Resistor—5 ohms, 5 watts
14671	Resistor—33 ohms, ½ watt
13428	Resistor—150 ohms, ½ watt
13998	Resistor—22,000 ohms, ½ watt
12454	Resistor—33,000 ohms, ½ watt
12412	Resistor—47,000 ohms, ½ watt
12264	Resistor—220,000 ohms, ½ watt
12285	Resistor—470,000 ohms, ½ watt
12679	Resistor—2.2 meg., ½ watt
13601	Resistor—10 meg., ½ watt
32945	Shaft—Tuning knob shaft and bushing

STOCK No.	DESCRIPTION
32969	Socket—Dial lamp socket
14278	Socket—Phonograph socket
32537	Socket—Tube socket
30585	Spring—Drive cord spring
33319	Transformer—First i-f transformer
32578	Volume control and power switch
	POWER LINE FILTER PLF-10
13057	Capacitor—60 mmfd.
12484	Capacitor—0.25 mfd.
33492	Coil—Choke coil
33493	Receptacle—Power receptacle
33491	Switch
	SPEAKER ASSEMBLIES (39105—2)
32964	Transformer—Output transformer
	MISCELLANEOUS ASSEMBLIES
X-639	Cabinet—Ivory finish—Model 5X5I......(net)
X-638	Cabinet—Walnut finish—Model 5X5W...(net)
32942	Dial—Glass dial scale
33317	Fastener—Push fastener to hold cabinet back
33306	Knob—Black tuning knob—Model 5X5I
32447	Knob—Ivory knob—Model 5X5W
32943	Nut—Speed nut to hold dial
31646	Spring—Knob retaining spring

Additional Replacement Parts:

Stock No.	
32946	Drum—Condenser drive drum and indicator
11765	Lamp—Dial lamp, Mazda No. 51
33324	Switch—"Remote" switch
32967	Transformer—Second I-F trans.
34569	Speaker—Complete—less transformer

A vacuum-tube radio schematic

But contemporary user manuals lack this depth of information. The most complex diagram I've seen in a Mac Pro user guide instructs you on how to sit at the computer: keep your "thighs tilted slightly," "shoulders relaxed," and so on.

What happened? Did electronics just get too hard and complex? On the contrary, improving electronics got too *easy*: the pace of Moore's law has been too much for small-scale innovators to keep up.

Where We Are: "Sit and Wait" vs. "Innovate"

Consider this snapshot of Moore's law, which states that "goodness" (pick virtually any metric: performance, transistor density, price per quanta, etc.) doubles every 18 months.

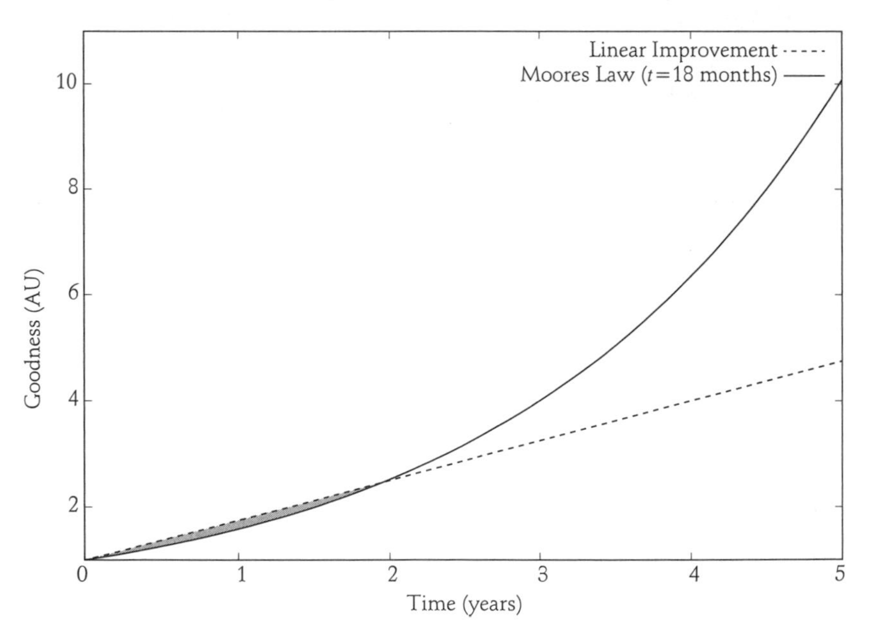

Moore's law, doubling once every 18 months versus linear improvement of 75 percent per year. The shaded sliver between the two lines at t < 2 *years represents the window of opportunity where linear improvement exceeds Moore's law.*

This chart is unusual in that the vertical axis is linear. Most charts depicting Moore's law use a logarithmic vertical scale, which flattens the curve's sharp upward trend into a much more innocuous-looking straight line. The shaded area, on the other hand, represents a linear improvement over time. This might represent a small innovator working at a constant, noncompounding, but respectable rate of 75 percent per year to add or improve features on a given platform. The tiny (almost invisible) space enclosed by the curves represents the market opportunity of the small innovator versus Moore's law.

The juxtaposition of these two curves highlights the central challenge facing small innovators. Sitting and waiting have long been more profitable than innovating. If it takes two years to double the performance of a system, you're better off simply waiting and upgrading to the latest hardware in two years. Racing against Moore's law is a Sisyphean exercise.

This exponential growth mechanic favors large businesses with the resources to achieve huge scale. Instead of developing one product at a time, a competitive business must have the resources and vision to develop three or four generations of products simultaneously. Reaching the global market within the timespan of a single technology generation requires a supply chain and distribution channel that can do millions of units a month: selling at a rate of 10,000 units per month, reaching "only" a million users, or about 1 percent of the households in the United States alone, would take eight years. And significantly, the small barrier (a few months' time) created by closing a design and forcing the competition to reverse-engineer products can be an advantage, especially against the pace of Moore's law.

Thus, technology markets have become inaccessible to small innovators as individuals struggle to keep up with the technology treadmill and big companies continue to close their designs to gain a thin edge on their competition. This trend is changing, however.

Where We're Going: Heirloom Laptops

Gordon Moore, the man who observed Moore's law, is one of Intel's co-founders. Moore's law is best known for describing how transistor density, and by extension CPU performance, would increase over time. For instance, consider this plot of Intel CPU clock speed at introduction versus time.*

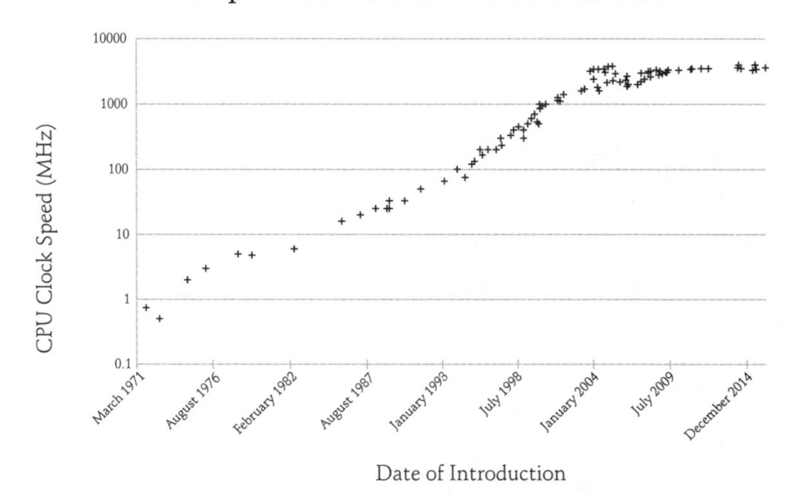

CPU clock speed over time. The plateau has held steady since 2014.

Notice the abrupt plateau where clock speed stops increasing. At that point, CPU makers started using multicore technology to drive performance, but this wasn't by choice. CPUs reached physical limits that prevented practical clock scaling, primarily related to power and wire delay scaling. Transistor density, and hence core count, continues to increase over time, but the pace is decelerating. Transistor count used to double once every 18 months; then it slowed down to double less than once every 24 months. Eventually, transistor density scaling will effectively end. The absolute endpoint for transistor scaling is a topic of debate, but one study† indicates that

---

* Data primarily from *https://en.wikipedia.org/wiki/List_of_Intel_microprocessors* and *https://en.wikipedia.org/wiki/List_of_Intel_Core_i7_microprocessors*. I track Intel CPUs because historically they have led the MHz curve and thus provide the most rigorous interpretation of Moore's law.

† H. Iwai, "Roadmap for 22nm and Veyond," *Microelectronic Engineering* 86, no. 7–9 (2009), doi: 10.1016/j.mee.2009.03.129.

scaling may stop at an effective gate length of about 5 nm. That's about the space between 10 silicon atoms, so even if this guess is wrong, it can't be wrong by much.

The implications are profound. One day, you won't be able to rely on buying a faster computer next year. Your phone won't get any smaller or more powerful. And the flash drive you buy next year will cost the same and store the same number of bits as the one you bought this year. The idea of an "heirloom laptop" may sound preposterous today, but someday, we may perceive our computers as cherished and useful heirlooms to hand down to our children as part of our legacy.

An Opportunity for Open Hardware

This slowing trend is good for small businesses, and likewise open hardware practices. To see why, let's revisit the plot of Moore's law versus linear improvement. This time, I'll overlay two new scenarios: technology doubling once every 24 and 36 months.

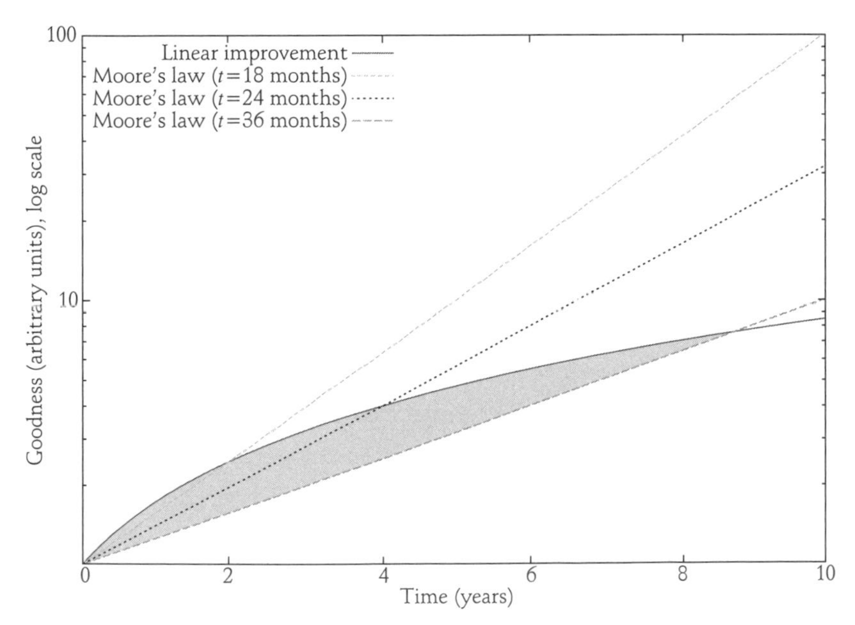

Three different Moore's law scenarios. The shaded sliver between linear improvement and the t=18 *months scenario turns into a large region of opportunity under the* t=36 *months scenario. (Note that the vertical axis is log scale.)*

The area bounded by the curved line and the straight line at the bottom represents the market opportunity for linear improvement versus Moore's law. In the 36-month scenario, not only does linear improvement have over eight years to go before it is lapped by Moore's law, but also there is a point at around year two or three where the optimized solution is clearly superior to Moore's law. In other words, there is a genuine market window for monetizing innovative solutions at a pace that small businesses can handle.

As Moore's law decelerates, there's also potential for greater standardization of platforms. Creating a standard tablet or mobile phone chassis with interchangeable components may seem ridiculous now, but it becomes a reasonable proposition when components stop shrinking and changing so much. As technology decelerates, there will be a convergence between hardware found in mobile phones and hardware found in embedded CPU modules like the Arduino. Just look at the Raspberry Pi, which was introduced in 2012. Models released in 2016 offer a quad-core, 1.2GHz CPU for performance comparable to entry-level smartphones at the time.

Creating stable, performance-competitive open platforms will empower small businesses. Of course, a small business can still choose to be closed, but by doing so, it must create a vertical set of proprietary infrastructure, and the dilution of focus to implement such a stack could be disadvantageous.

In the post–Moore's law future, FPGAs may perform respectably compared to their hardwired CPU kin, for at least two reasons. First, the flexible yet regular structure of an FPGA may lend it a longer scaling curve, in part due to the FPGA's ability to reconfigure circuits around small-scale fluctuations in fabrication tolerances. Second, the extra effort to optimize code for hardware acceleration will amortize more favorably as CPU performance scaling increasingly relies upon difficult techniques like using parallel cores on a massive scale. Massively multicore CPU architectures look a lot like the coarse-grain FPGA architectures proposed in academic circles in the '90s. An equalization of FPGA-to-CPU performance should greatly facilitate the penetration of open hardware at a deep level.

There will be a rise in repair culture as technology becomes less disposable and more permanent. Replacing worn-out computer parts five years from their purchase date won't seem so silly when the replacement part has virtually the same specifications and price as the old part. This rise in repair culture will create a demand for schematics and spare parts that in turn facilitates the growth of open ecosystems and small businesses.

Personally, I'm looking forward to the return of artisan engineering, where elegance, optimization, and balance are valued over feature creep, and where I can use the same tool for a decade and not be viewed as an anachronism. (Most people laugh when they hear I held on to Eudora 7 as my email client until 2012, when I switched to my current client, Thunderbird.)

The deceleration of Moore's law has already impacted markets that are less sensitive to performance. Consider the rise of Arduino. It took several years to gain popularity, with virtually the same hardware at its core the whole time. Fortunately, the demands of Arduino's primary market (physical computing, education, and embedded control applications) have not grown, allowing the platform to remain stable. This stability has enabled Arduino to grow deep roots in a thriving user community with open and interoperable standards.

With some hard work and a bit of luck, I believe the open hardware ecosystem will surely blossom. The inevitable slowdown of Moore's law may spell trouble for technology giants, but it will also create an opportunity for the open hardware movement to grow roots and start something potentially very big. To seize this opportunity, open hardware pioneers will need to set the stage by creating a culture of permissive standards and customs that can scale over time.

I look forward to being a part of open hardware's bright future.

CLOSING THOUGHTS

Although chumby, conceived in 2006, was a bit ahead of its time and the company ultimately fell victim to Moore's law, my reflections on the slowing pace of Moore's law encouraged me to try yet another experiment in open hardware. The next chapter, on Novena, shares the story of my quixotic adventures building a bespoke open source laptop.

7. novena: building my own laptop

It was 2012, and I was unemployed. My previous startup had failed, and I was taking a year off to figure out what I should do next. My friend xobs (introduced in Chapter 4) and I had a tradition that we maintain to this day: every Friday, we sit down for a few beers at lunch and shoot the breeze. During one of those "Beer Friday" discussions, we decided to build our own laptop. I expressed displeasure with how I'd never been employed to build a product that I'd actually want to use every day. As a design engineer, you're typically driven by market requirements, not your own eclectic tastes. We bantered a bit about things we'd find useful and realized that, thanks to the gradual slowing of Moore's law, maybe it wasn't so crazy for us to build an open laptop with some wacky features just for hackers. From there, we started a hobby project to build a computer just for ourselves, something we'd use every day that would be easy to extend and mod—our very own electronic Swiss Army knife. We gave the project the code name Novena, the name of a Singaporean metro station and Latin for "nine."

The second-generation Novena design that went up on Crowd Supply

The finished Novena was a 1.2GHz, Freescale (now NXP) i.MX6 quad-core ARM architecture computer closely coupled with a Xilinx FPGA. It was designed for users who wanted to modify and extend their hardware: all the documentation and PCBs were and still are open and free to download,* and we gave it a variety of features that facilitated rapid prototyping.

NOT A LAPTOP FOR THE FAINT OF HEART

As I talked to more people about Novena, however, I realized that others were interested in owning a laptop like that but perhaps didn't want (or didn't know how) to make their own circuit boards. In response to the overwhelmingly positive feedback we received to a blog post on the topic, xobs and I launched a campaign on Crowd Supply in 2014, once the design was stable and tested. Over 1,000 people pledged their support; I am happy to report that we fulfilled every single campaign pledge, most of them within a few months of the promised date. After the campaign's close, we decided it would spread our limited resources too thin to maintain the supply chain for the full laptop configuration, but we would sell and support the Novena motherboard hardware for at least five years from the launch of the campaign.

To be clear, Novena is not a machine for the faint of heart. It's an open source project, which means part of the joy (and frustration) of the device is that it is continuously improving. It's perhaps the only laptop that's ever shipped with a screwdriver. Anyone who bought one of the original designs had to install the battery and screw on the LCD bezel of their choice—green or blue. The speakers came as a kit so users wouldn't have to use our speaker box design. If someone had access to a 3D printer, they could make and fine-tune their own speaker box.

* You can find the documentation online via the Kosagi wiki at *http://www.kosagi.com/*.

Despite all of those DIY options, I wasn't looking to break any low-price records with Novena. It was designed as a low-volume, handcrafted laptop made with uniquely open source components, and the cost matched the design. We offered three tiers:

- An "all-in-one desktop" option for $1,195 that was ready to use with a keyboard and mouse out of the gate, but needed to be plugged in
- A "laptop" option for $1,995 that included a battery controller board, for hackers on the go
- An "heirloom laptop" tier for $5,000 that came in a gorgeous, handcrafted wood-and-aluminum case

In Chapter 6, I said that as Moore's law slows down, I predict parents passing down computers to their children. The Heirloom Novena is meant to be treated that way, though it has the same hardware on the inside as the other two options.

But those prices weren't so different from the prices of high-end consumer laptops. The biggest challenge was figuring out how to offer something so custom and complex at that price point, in low volumes. We weren't looking to recover the research and development cost in the campaign; that's a sunk cost, as anyone is free to download the source and benefit from our thoroughly vetted design today. Our minimum funding goal of $250,000 was a tiny fraction of what's typically required to recover the million-dollar-plus investment behind the development and manufacture of a conventional laptop; xobs and I met this challenge with a combination of know-how, unique design, and strong relationships with our supply chain.

DESIGNING THE EARLY NOVENA

We optimized the Novena's design to reduce the amount of expensive tooling required, while still preserving our primary goal of it being easy to hack and modify. We spent a year and a half poring over three revisions of the PCBA until we were confident that the complex design would be functional and producible. We also optimized certain tricky components, such as the LCD and the internal display port adapter, for reliable sourcing at low volumes. Finally, I spent a few months traveling the world, lining up a supply chain that could deliver this design (even in low volume) at a price comparable to other premium laptops.

Of course, all the design documentation is open, so with sufficient skill and resources, you could build a Novena from scratch yourself. I chose the hardware and its subcomponents to make this the most practically open hardware laptop I could with state-of-the-art technology. You can download, without NDA, the datasheets for all the components, and key peripheral options were chosen such that you can build a complete firmware from source with no opaque blobs.

Under the Hood

This board's dimensions are approximately 121 mm × 150 mm; it's sized to fit comfortably underneath a standard-sized laptop keyboard (though the image is rotated compared to the installation orientation). As you can see in the full laptop photos earlier in the chapter, the port farm is on the right side of the laptop, not the bottom. The board is just under 14 mm thick, a height set by the thickness of an Ethernet connector. The base portion of my Lenovo T520 is just under 24 mm thick, and once a keyboard and plastics are stacked on this board, the base of the Novena comes to just about the same thickness.

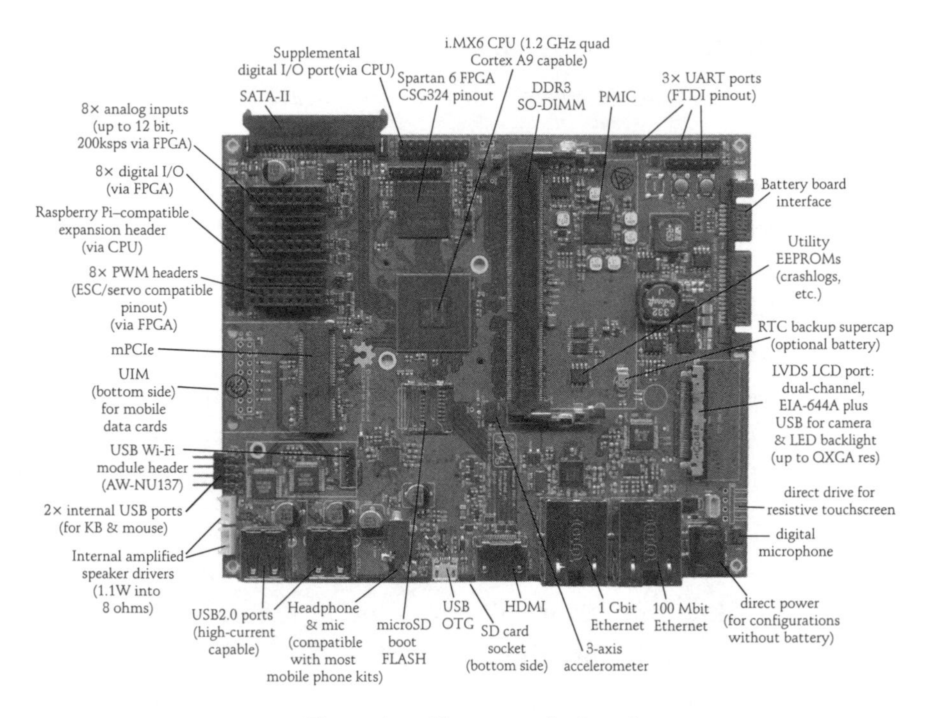

The earliest Novena motherboard

Now let's look at some of the motherboard's features.

PRELIMINARY FEATURES

The first iteration of the Novena motherboard used a Freescale iMX6 CPU, which has an NDA-free datasheet and programming manual. In the lists that follow, items marked with a double asterisk (**) require a closed-source firmware blob, but the system is bootable and usable without the blob.

The CPU footprint we used could support the following quad- and dual-lite versions of the iMX6:

- Quad-core Cortex A9 CPU with NEON FPU @ 1.2 GHz
- Vivante GC2000 OpenGL ES2.0 GPU, 200Mtri/s, 1Gpix/s**

This version of Novena booted from microSD firmware. In terms of other internal memory, it had a 64-bit, DDR3-1066 SO-DIMM, which could be upgraded to 4GB, and a SATA-II (3Gbps) hard drive.

Novena was full of internal ports and sensors from the start, too. These are the highlights:

- A Mini PCI-express (mPCIe) slot, for blob-free Wi-Fi, Bluetooth, mobile data, and so on
- A UIM slot, for mPCIe mobile data cards
- A dual-channel LVDS LCD connector with up to QXGA resolution (2,048 × 1,536 px) at 60 Hz and a USB 2.0 side channel for a display-side camera
- A resistive touchscreen controller (capacitive touch displays, on the other hand, typically come with an integrated controller)
- 1.1 W, 8-ohm internal speaker connectors
- Two USB2.0 internal connectors, for a keyboard and mouse or trackpad
- A digital microphone
- A three-axis accelerometer
- A header for an optional AW-NU137 Wi-Fi module**

We made the following ports externally accessible:

- HDMI
- The SD card reader
- The headphone and microphone jacks (compatible with most mobile phone headsets, these also supported sensing inline cable buttons)
- Two USB 2.0 ports, supporting high-current (1.5A) device charging
- A 1Gb Ethernet port

And, of course, since xobs and I were making the Novena for ourselves, we included a bunch of other "fun" features that we knew would be great for hackers:

- 100Mb Ethernet (dual Ethernet capability allows Novena to be used as an inline packet filter or router)
- USB On-the-Go (enables the Novena to spoof or fuzz Ethernet, serial, and other connections over USB via a gadget interface to other USB hosts)
- A utility serial EEPROM, for storing crash logs and other bits of handy data
- A Spartan-6 CSG324-packaged FPGA with several interfaces to the CPU, including a 2Gbps (peak) RAM-like bus—for bitcoin mining, or whatever else you might want to toss in an FPGA
- Eight FPGA-driven 12-bit, 200ksps analog inputs
- Eight FPGA-driven digital I/O pins
- Eight FPGA-driven PWM headers, compatible with hobby ESC and PWM pinouts (enables direct interfacing with various RC motor/servo configurations and quad-copter controllers)
- Raspberry Pi–compatible expansion header
- Thirteen CPU-driven supplemental digital I/Os
- Three internal UART ports

We tweaked those specs going into production, making the most drastic changes around the FPGA expansion connectors. Instead of a cluster of motion-control-focused headers, we opted to install a header capable of high data rates, which xobs and I used to great effect in future projects involving the Novena.

THE BATTERY BOARD

To give maximum power management flexibility, I implemented the battery interface functions on a daughtercard. I co-opted a cheap and common SATA-style connector to route power and control signals between the mainboard and the daughtercard. To prevent users from accidentally plugging a hard drive into the battery port, I inverted the gender of the battery-SATA connector from the actual mass-storage SATA-II connector.

The battery card in the first Novena board was meant to work with the battery packs used by most RC enthusiasts: LiPo packs ranging from 2S1P to 4S1P (that is, two-cell to four-cell). RC packs are great because they're designed for super-fast charging and they're cheap and easy to buy. For the board-side battery plug, I decided to use the Molex connector found on classic disk drives, since they are cheap, common, and easy to assemble with simple tools. I couldn't use a standard RC connector because the vast majority of them are designed for inline use, and the few that have board mounts were too thick or too weird for this application.

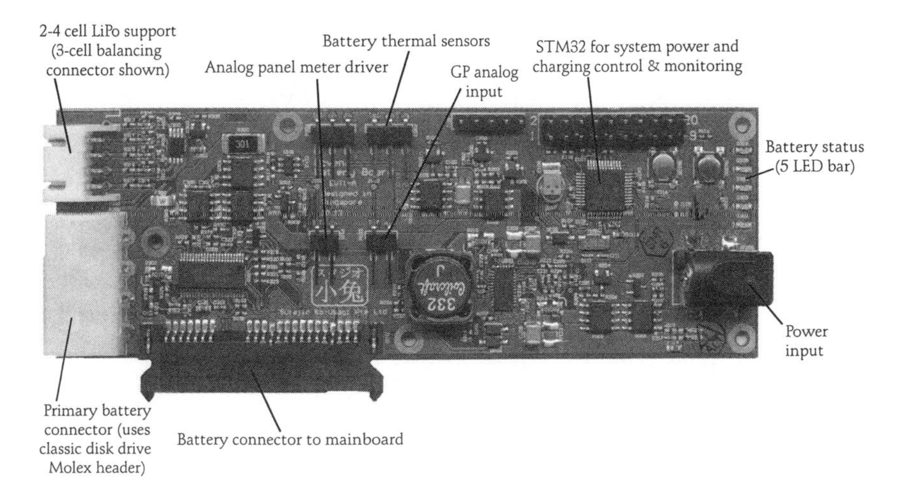

The preliminary Novena battery board

The battery board could charge batteries at rates in excess of 4A; for example, charging a three-cell, 45 Wh (4 Ah) battery took about one hour. If typical power consumption were around 5 to 6 W per hour, that would be seven or eight hours of runtime with a one-hour charge time. Of course, since the whole laptop was user-configurable, typical power consumption was really hard to estimate. If a user dropped in a monster LCD and a power-hungry magnetic hard drive with loads of peripherals, the power consumption would be much higher.

xobs suggested another cute power-related feature that made it into the design. He thought it would be neat to embed a retro analog needle meter into the palm rest of the laptop to display power consumption in real time. I thought it was a great idea, so I designed that into the circuit board. Of course, the analog meter is driven by a DAC on the battery microcontroller, so it could be configured to perform a multitude of useful (or not so useful) analog readouts, such as remaining runtime, battery voltage, temperature, the time (represented as an analog value), and so on.

After spending a couple of months validating all the features (it was a long list of features to grind through), we ported drivers and a Linux distro to the board. That was no small task either, but thankfully, I had xobs's skillful help, and we got the job done.

The Enclosure

From there, I was really looking forward to designing the enclosure. For the first revision, I thought about making something out of laser-cut acrylic that would be vaguely tablet-like, to avoid having to mess around with a friction clutch on the first go at a case. I ended up hand-building our first prototype cases from aluminum and leather, to validate the laptop use case for Novena. That design was rough; as Cory Doctorow put it on *Boing Boing*, it was "gloriously fuggly."*

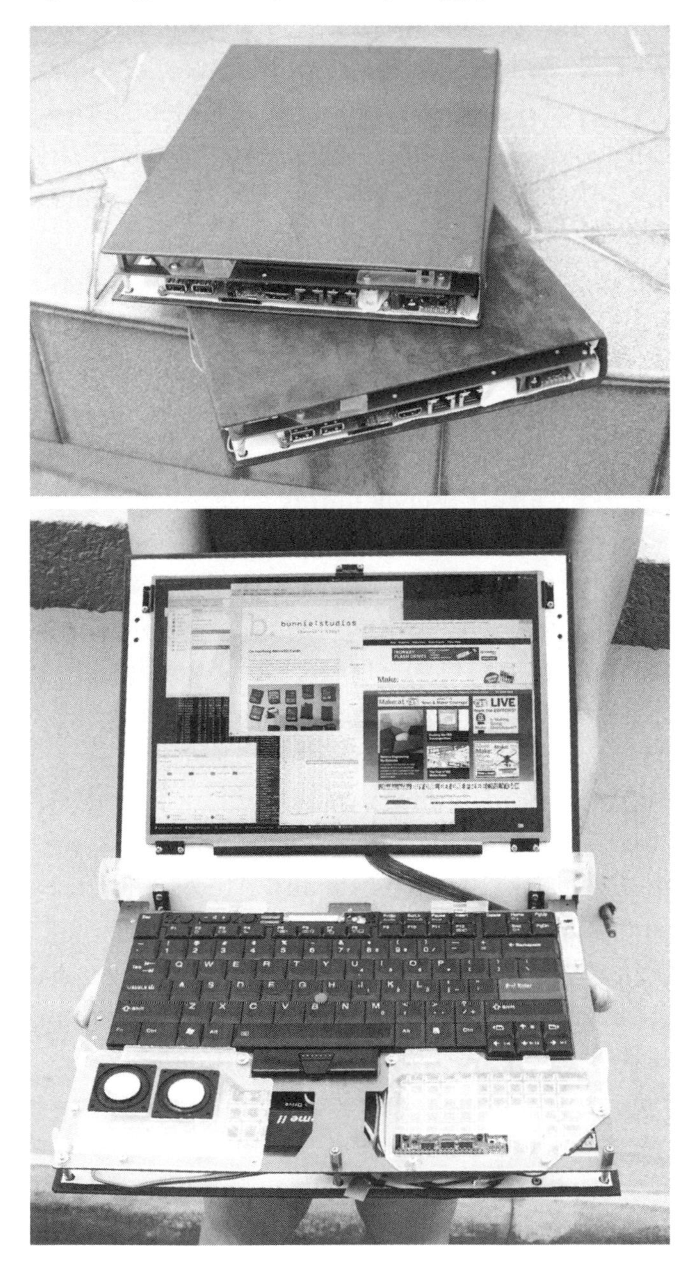

I love that my laptop smells of leather when it runs!

---

* *http://boingboing.net/2014/01/17/building-a-fully-open-transpa.html*

The second-generation Novena case I showed earlier is sleeker. The first thing you probably noticed about the design is that it opens the "wrong" way. This feature allows the Novena to be used as a wall-hanging unit when the screen is closed. It also solves a major problem I had with the original clamshell prototype: it was a real pain to access the hardware for hacking, as it was blocked by the keyboard mounting plate.

In the version we sold on Crowd Supply, the screen automatically pops open with the slide of a latch, thanks to an internal gas spring. (Novena isn't just an open laptop—it's a self-opening laptop!) We intentionally left the internals naked in this mode for easy access, but bare internals also make clear that Novena isn't for casual home users.

We included an array of mounting bosses—which we called a *Peek array*—as well, to facilitate hackability. Normally, laptops have mounting points only for the handful of features designed into their original blueprints. But a hackable laptop must accommodate a huge space of possible peripherals. Instead of requiring users to drill holes or glue things down in their laptop cases, we provided a regular array of threaded inserts. It was a bit like a breadboard, but for rapid mechanical prototyping. To help define the array, I consulted with Nadya Peek, a graduate student at MIT's Center for Bits and Atoms and an expert in digital fabrication—hence the name *Peek array*.

Another feature of the second-generation design is that the LCD bezel is made of a single, simple aluminum sheet. This allows anyone with access to a minimal machine shop to modify or craft their own bezels; no custom tooling required. My hope with that design was to make adding knobs and connectors or changing the LCD relatively easy for Novena hackers. To encourage users to experiment, we shipped desktop and laptop Novenas with two LCD bezels so no one had to worry about having an unusable machine if they messed one up while experimenting.

Most laptops have a keyboard and mouse attached to the enclosure, but the Novena has a detached keyboard and trackpoint because that feature was attractive to me personally. I'd always wanted a display I could "hang" on the seat in front of mine when sitting in an airplane or a bus: it's a lot easier on the neck, and the arrangement actually works *better* if the person in front reclines their seat.

While I was still considering whether to do a clamshell design or some other funky design for the exterior, I also thought about trying an enclosure made of wood and brass. After all, the whole idea of making my own laptop was to play around with some new ideas! As mentioned earlier, we actually did wind up doing a limited run of a wooden-cased Novena that we dubbed the *heirloom laptop*.

The Heirloom Novena laptop

THE HEIRLOOM LAPTOP'S CUSTOM WOOD COMPOSITE

When mainline Novena production was finally humming along in April 2015, I spent a week in Portland, Oregon, working alongside Kurt Mottweiler (a designer and woodworker who specializes in making cameras with wooden enclosures) to hammer out all of the final open issues on the Heirloom devices. xobs and I are certainly proud of how the Heirloom Novenas turned out!

Working with Kurt on the Heirloom laptop

Growing Novenas

In a literal sense, the Heirloom Novenas were "grown." Wooden enclosures meant important structural elements came from trees. Making every laptop identical would have been easy, but we felt it would be much more apropos of a bespoke product to make each laptop unique by picking the finest woods and matching their finish and color in a tasteful fashion. As a result, no two Heirloom laptops look the same; each is uniquely beautiful.

Some handpicked wood, waiting to become a Novena case

A lot of science and engineering went into the Heirloom laptops, too. For starters, Kurt created a unique composite material by layering cork, fiberglass, and wood. To help characterize the novel composite, we took some material samples to the Center for Bits and Atoms, where Nadya Peek and Will Langford characterized the performance of the material. We took sections of the wood composite and performed a three-point bend test using an Instron 4411 electromechanical material testing machine.

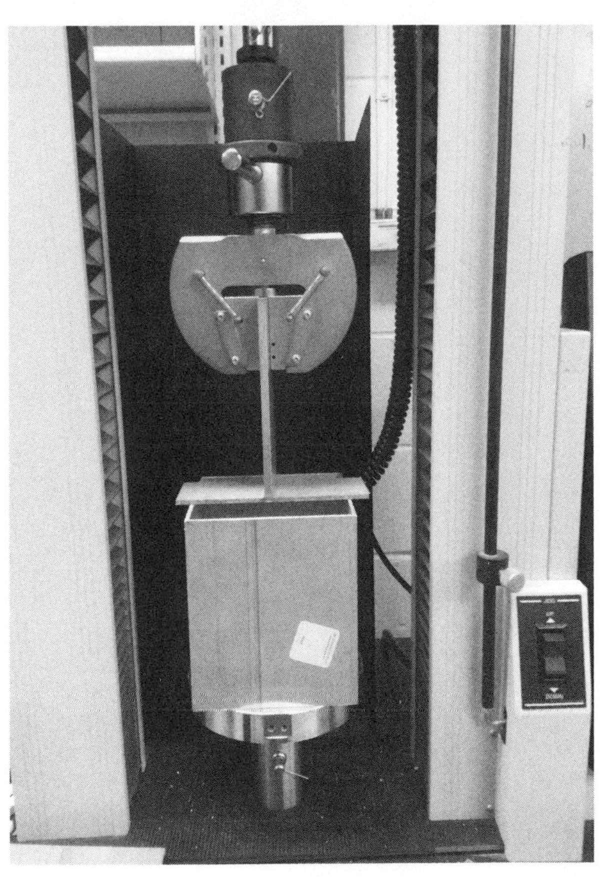

Heirloom composite material loaded into the testing machine

The Mechanical Engineering Details

From the test data, we were able to extract the flexural modulus (also called Young's modulus) and flexural strength of the material. I'm not a mechanical engineer by training, so terms like *modulus* and *specific strength* kind of go over my head. But Nadya was kind enough to lend me some insight. She pointed me at the Ashby chart, which, as with some xkcd comic panels, I could stare at for an hour and still not absorb all the information contained within.

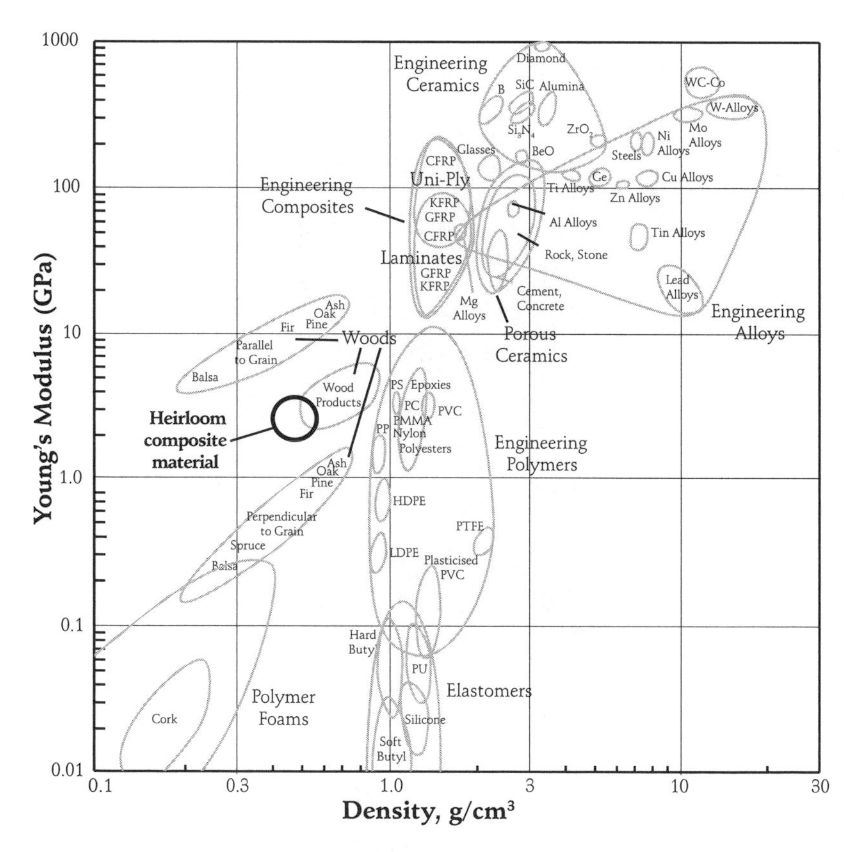

The Ashby chart plots Young's modulus versus density for many materials. The annotated area shows approximately where the Heirloom composite material lands.

The bottom left of the chart shows bendy, light materials like cork, and the top right of the chart has rigid, heavy materials, like tungsten (W). For a laptop case, we wanted a material with the density of cork but the stiffness of plastic. Wood products occupy a space in the chart to the left of plastics, meaning they are less dense, but they have a problem:

they are weak perpendicular to the grain. Depending on the direction of the strain, wood can be as yielding as polyethylene (the material used to make plastic shopping bags) or stiffer than polycarbonate (the material layered with glass to make bulletproof windows). Composite materials are great because they allow you to blend the characteristics of multiple materials to hit the desired characteristic. In the Heirloom laptop's case, Kurt blended cork, glass fiber, and wood.

The measurements of the Heirloom composite show a flexural strength of about 33 megapascals, and a flexural modulus of about 2.2 to 3.2 gigapascals.* The density of the material is 0.49 g/cm^3, meaning it's about half the density of ABS plastic, the plastic LEGO bricks are made from. As shown on the Ashby chart, plotting these numbers reveals that the Heirloom composite occupies a nice spot to the left of plastics and provides a compromise on stiffness based on grain direction. And during testing, the material didn't fail catastrophically.

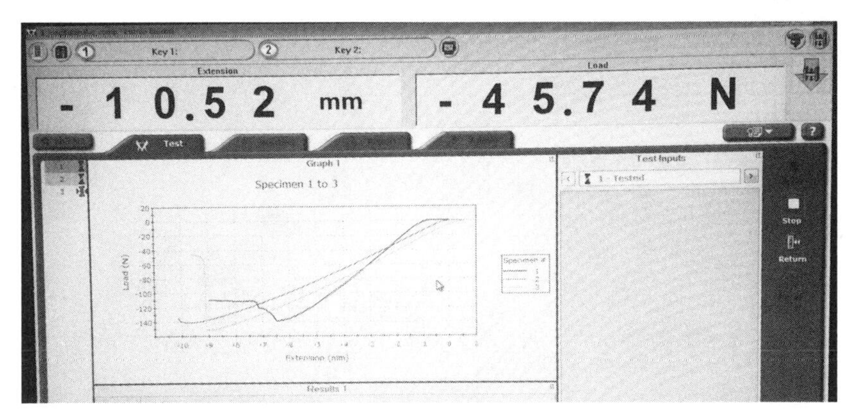

Graphs of load versus extension on the Heirloom laptop composite, as plotted by the Instron testing machine

Even after being bent past its peak load, the composite was still mostly intact and providing resistance. This result was a bit surprising. We had expected the material to break in two

* One megapascal is 1 newton (unit of force) per mm^2; 1 gigapascal is 1 kilonewton per mm^2.

on failure, like natural wood. Furthermore, after we reset the test, the material bounced back to its original shape. We bent the composite by over 10 mm, but once the load was removed, I could barely tell it went through testing. This high fracture toughness and resilience are desirable properties for a laptop case.

Of course, watching a machine go to work on the material was fun, but there's nothing quite like holding it yourself. I still remember picking up the material, feeling how light it was, giving it a good bend, and being surprised by its rigidity and ruggedness.

CHANGES TO THE FINISHED PRODUCT

From the moment Novena was successfully crowdfunded, an incredible team of people worked to make it a reality. With help from the engineers and product managers at our manufacturing partner, AQS, Novena's case moved from prototype to pilot production just four months after the campaign.

The conference room where we did the T1 plastics review in Dongguan, China

Sure, xobs and I did plenty of work on our own before we even started the crowdfunding, but it takes many hands to build a product of this complexity. We couldn't have done it without our dedicated and hardworking team at AQS. I've said before that your factory is your partner, and thanks to a great partner, we were able to get this done in a short amount of time.

Case Construction and Injection-Molding Problems

By the late summer of 2014, the Novena cases we were carrying around were made of entirely production-process hardware—no more hand-built prototypes. To get there, we'd opened a total of 10 injection-molding tools; for comparison, a product like NeTV or chumby had perhaps 3 or 4 injection-molding tools.

As I briefly described in Chapter 1, injection molding is a process where plastic is molded into a net shape. Hot, high-pressure liquid plastic is forced into a hardened steel cavity called a *tool*. The steel tool is a masterpiece of engineering in itself: it's a water-cooled block weighing about a ton and capable of handling pressures found at the bottom of the Mariana Trench, and the internal surfaces are machined to tolerances better than the width of a human hair. On top of that, the tool contains a clockwork of moving pieces, with dozens of ejector pins, sliders, lifters, and parting surfaces that come apart and back together again smoothly over thousands of cycles. It's amazing that tools of such complexity and refinement can be crafted in a couple of months.

With so many moving parts, it's no surprise that the tools required several iterations of refinement to get absolutely perfect. In tooling jargon, the iterations are referred to as T0, T1, T2, and so on. You're doing pretty well if you can go to full production at T2; thankfully, our T1 plastics were 99 percent of the way there, meaning we had an easy path to full production. T1 had just a few issues relating to flow and knit lines, as well as spots where the plastic warped during cooling or bound itself to the tool during ejection, causing deformation. This manifested itself as spots where the seams weren't as tight as we wanted them to be in the case, and with just a little bit of tuning, we were production-ready.

Most people have only seen products of finished tooling, so I'll share what a pretty typical T0 (first-attempt) shot looks like, particularly for a large and complex tool like the Novena case base part. Test shots like this are typically done with scrap resin in light colors that highlight defects. We used gray plastic here to make tuning the mold easier, but the final units had black bases.

Some T0 shots of the base of the Novena case. The regular array of circles on the left in the top photo form the basis of the Peek array. To make the array, threaded brass inserts were heat-staked into the circular bosses after injection molding.

There's a lot going on with this piece of plastic. Let's zoom in on some of the artifacts.

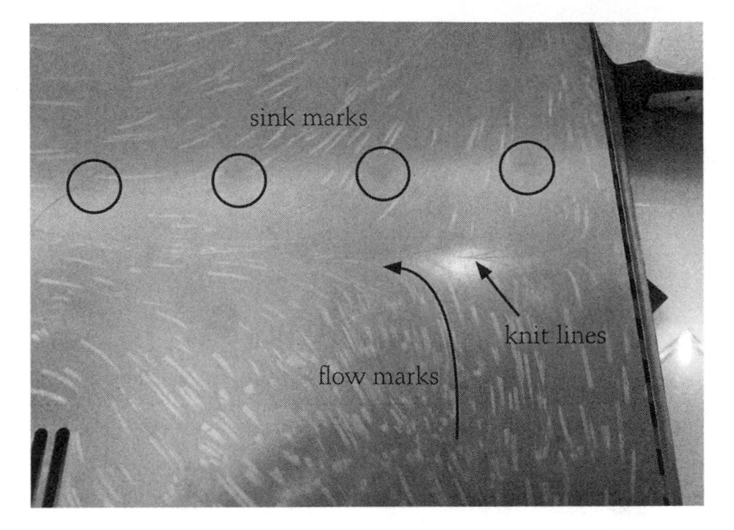

A visual guide to the deformations in the T0 case base

The circles highlight a set of *sink marks*, which happen when the opposite side of the plastic has a particularly thin or thick feature. These areas cool faster or slower than the bulk of the plastic, causing them to pucker slightly and create a sort of shadow. Sink marks are particularly noticeable on mirror-finish parts. In this case, the sink marks happened because the plastic underneath the nut bosses of the Peek array were much thinner than the surrounding plastic. To fix this problem, we thickened that region slightly, reducing the overall internal clearance of the case by 0.8 mm. That was possible because fortunately, I'd designed the case with a little extra clearance margin.

The straight arrow points to a *knit line*. This is a region where plastic flow meets within the tool. As plastic is injected into the cavity, it tends to flow from one or more gates, and where the molten plastic meets itself, a hairline scar forms. Knit lines are often located at points of symmetry between the gates where the plastic is injected. On this tool, there were four gates located underneath the spot where the rubber feet go. Gates are considered cosmetically unattractive, and thus we placed them strategically to hide their location.

The white feathery artifacts indicated by the curved arrow are *flow marks*. These streaks appeared because the plastic cooled a bit too quickly within the tool. You can often fix this problem by adjusting the injection pressure, cycle length, and temperature. It's best to use test shots on the molding machine to make those tweaks. You can tweak one parameter at a time, shot after shot, until you find an optimum cooling speed. This process can sometimes take hundreds of shots, creating a small hill of scrap plastic as a by-product.

Most of these gross defects were fixed by T1, and at that point, the plastic looked much closer to production-grade. We were also able to start using black-colored plastic, which tends to hide defects.

There were still a few issues around fit and finish, of course. But despite them, the case felt much more solid than the prototypes, and the gas piston mechanism was finally consistent and really smooth.

The T1 case base, in initial testing after the live hardware was transferred into the plastics

Changes to the Front Bezel

The front bezel of Novena's case (not to be confused with the aluminum LCD bezel) went through some changes after the campaign. When we closed funding, it had two outward-facing USB ports and one switch. Novena shipped with two switches, one outward-facing USB port, and one inward-facing USB port.

One switch is for power: it goes directly to the power board and can be used to turn the system on and off even when the main board is fully powered down. The other switch is wired to a user keypress to facilitate Bluetooth association for keyboards that are being stupid. Some keyboards can take up to a half-minute to cycle through *something* (presumably, it's security-related) before they connect. There are hacks for bypassing that, but you'd have to run a script on the host. Our idea was that by pressing this button, users could trigger a convenience script to get past the utter folly of Bluetooth. This switch also doubles as a wake-up button for when the system is suspended.

As for the USB ports, the design still had four in total, but the configuration became as follows:

- Two higher-current-capable ports on the right
- One standard-current-capable port on the front
- One standard-current-capable port facing toward the Peek array

In other words, we faced one USB port toward the inside of the machine. Since half the fun of Novena is modding the hardware, I figured a USB port on the inside would be at least as useful as one on the outside.

For users who wouldn't do hardware mods, an inside USB port would also be a fine place to plug small dongles that generally stay attached, like the radio transceiver for a keyboard. It's a little inconvenient to initially plug in the dongle, but keeping the radio transceiver dongle facing inside helps protect it from damage when you throw your laptop into your travel bag.

DIY Speakers

We toyed with several speaker options for Novena. A core idea behind the design was to encourage every user to choose their own speaker. Some people really listen to music on their laptop when they travel, but others simply rely upon the speaker for notification tones and would prefer to use headphones for media capabilities. Physics dictates that high-quality sound requires a certain amount of space and mass. We wanted users with a more relaxed fidelity requirement to be able to reclaim the space and weight that nicer speakers would require.

Kurt Mottweiler selected a nice but very compact off-the-shelf speaker, the PUI ASE06008MR-LW150-R, for the Heirloom. When we found that the same speaker fit well into the standard Novena's Peek array and had acceptable fidelity, particularly for its size, we adopted it as the standard offering for audio. But we shipped it with a mounting kit for easy removal, so users who might need to reclaim the space (or who wanted to put in larger speakers) could do so with ease.

The PVT2 Mainboard

The Novena mainboard went through a minor revision prior to mass production. The fourth and final revision of the motherboard was known as the "PVT2" version. The majority of the changes focused on replacing or updating components that were at risk of reaching end-of-life. The two most significant additions from a design standpoint were an internal flexible printed circuit (FPC) header to connect to the front bezel cluster, and a dedicated hardware real-time clock (RTC) module.

We added the internal FPC header to improve signal routing from the mainboard to the front bezel cluster. We had to run two USB ports plus a smattering of GPIOs and power to the front bezel, and the original connection scheme required multiple cables. The updated design condensed that into a single FPC to simplify the design and improve reliability.

We included a dedicated hardware RTC module because the i.MX6's built-in RTC didn't perform well. The CPU simply had a higher leakage on the RTC than reported in the datasheet, and the lifetime of the RTC when the system was turned off was measured in, at best, minutes. We decided that there was too much risk in continuing to develop with the on-board RTC and opted for an external, dedicated RTC module that we knew worked. To increase compatibility with other i.MX6 platforms, we picked the same module used by the Solid-Run Hummingboard, the NXP PCF8523T/1.

It's also important to note that we completely overhauled the FPGA expansion header on our second revision of the motherboard. The version of the motherboard shown at the beginning of this chapter contained a cluster of headers optimized for motion control applications. We decided that our motherboard was too large for anyone to put it inside a quad copter, and perhaps the FPGA would see more use as a high-speed data acquisition and processing device. To enable this functionality, we gave the FPGA a dedicated 256MB of DDR3 memory and broke out high-speed differential signals to a connector capable of passing signals at rates exceeding a gigabit per second. Users could still use the FPGA for motion control applications, but they'd need to plug in a simple breakout board (like the GPBB I discuss next) to route our signals to the connector formats commonly used by motion control systems.

The updated Novena motherboard

A Breakout Board for Beginners

One of the rewards every backer received as thanks for supporting our campaign was a breakout board that we referred to as the *GPBB*, or the *General-Purpose Breakout Board*. Redesigning our FPGA expansion header on Novena to target high-speed applications also made getting started with the device much more difficult for entry-level hackers. Due to the constraints of physics, high-speed connectors tend to have very dense pin arrangements that are unfriendly to beginners. We designed the GPBB to help entry-level users work with the FPGA. The GPBB converts the dense, high-speed signal header on the FPGA into a beginner-friendly 0.1-inch-pitch, 40-pin header and includes a few LEDs and analog data converters to boot.

The final production GPBB

One growing challenge for beginners is the fact that Moore's law keeps on pushing down the allowable voltage range of digital I/Os. Newer generations of transistors run at lower voltages, which make them incompatible with the venerable +5 V standards most entry-level projects use. For instance, our FPGA could only handle signals up to +3.3 V. As a result, we built voltage translators into the GPBB that could safely handle +5 V and bring them down to the +3.3 V levels accepted by the FPGA.

The final version of the GPBB included a tweak enabling users to adjust the I/O voltage, instead of fixing it at +5 V. We provided a software setting to allow users to choose whether the GPBB's external I/Os default to 5 V or 3.3 V, and we designed the board so that users could adjust the lower voltage to 2.5 V or 1.8 V by changing a single resistor (R12). I labeled that resistor "I/O VOLTAGE SET" and made it a 1206 part, so soldering novices could make the change themselves.

The Desktop Novena's Power Pass-Through Board

The "all-in-one desktop" tier originally included just the desktop case, the Novena mainboard, and the front panel breakout. But that configuration made power management awkward, as I designed the overall power management system for the case assuming there would be a helper microcontroller managing a master cutoff switch.

Complexity is the devil, and getting the software going for even a single configuration was hard enough on its own. Ultimately, we found it cheaper to introduce a new piece of hardware to the power management system for the desktop, rather than deal with multiple code configurations.

Therefore, desktop systems shipped with a power pass-through board. It was a simple PCB assembly containing just the STM32 controller and power switch of the full battery board. This allowed us to use a consistent gross power management architecture across both the desktop and the laptop systems.

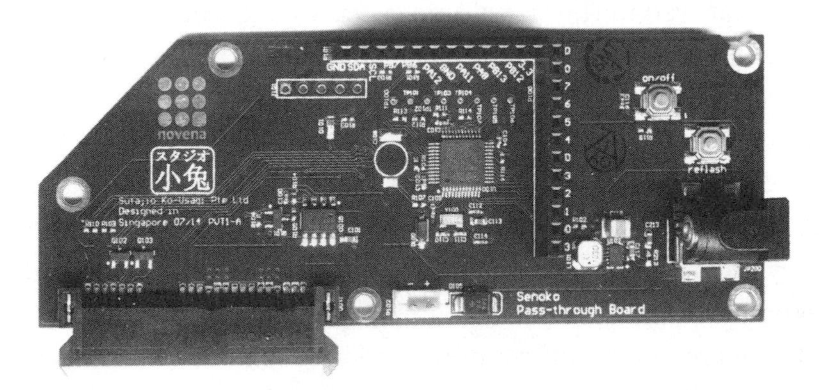

The desktop's pass-through board

This approach was like swatting a fly with a sledgehammer—but the sledgehammer cost as much as the flyswatter. Plus it's inconvenient to carry both a flyswatter and a sledgehammer around. So, yes, we used a 32-bit ARM CPU to read the state of a pushbutton and flip a GPIO, and yes, a full multithreaded real-time operating system (ChibiOS) ran underneath it all.

It did feel a little silly, though. That's why we broke out some of the unused GPIO pins, making Novena even more hackable. Hopefully, some clever user will find an application for all that untapped power!

Custom Battery Pack Problems

The battery pack for Novena was definitely a wildcard in the project stack. Building Novena was the first time xobs or I had made a system with such a high-capacity battery, and working through all the shipping regulations to get them delivered to customers was a challenge.

Some countries have particularly strict regulations around importing lithium batteries. In the worst case, we had to send some customers a laptop with no battery inside, and we shipped an off-the-shelf battery pack from a vendor that specializes in RC battery packs (like Hobby King) separately to those customers at our own cost. They got the same battery featured in the crowdfunding campaign, but they had to plug it in themselves. That was our safest fallback solution, since Hobby King ships thousands of battery packs a day all around the world.

Shipping woes didn't stop us from developing a custom battery pack, though. Maintaining a standing stock of battery packs is difficult because batteries need to be periodically conditioned, so only campaign backers got that battery pack—provided their country of residence allowed its import. We couldn't know for sure until we tried, but we did get UN38.3 certification for the custom battery pack. In theory, that certification would allow us to ship the batteries by air freight, but regulations around battery shipment are always in flux. It seems countries and carriers keep inventing new rules, particularly with all the paranoia about the potential use of lithium batteries as incendiary devices, and we didn't have the resources to keep up with the zeitgeist.

The custom pack's capacity was rated at 5,000 mAh, which is about twice the capacity of the pack we featured in the crowdfunding campaign. (That one had 3,000 mAh printed on the outside but delivered about 2,500 mAh in practice.) In real-life testing, the custom pack provided about six or seven hours of runtime with minimal power management enabled. Also, since I got to specify the battery, I knew it had the correct protection circuitry built into it and the provenance of its cells, so I was confident in its long-term performance and stability.

Choosing a Hard Drive

The crowdfunding campaign referenced providing 240GiB Intel 530 (or equivalent) and 480GiB Intel 720 drives for the laptop and heirloom models, respectively. We left the spec slightly ambiguous because the SSD market moves quickly. We knew the best drive when we drew up the spec would probably be different from the best drive we could get when we actually did the purchasing.

After doing some research, we felt the best equivalent drives at purchase time were the 240GiB Samsung 840 EVO (for the laptop model) and the 512GiB Samsung 850 Pro (for the Heirloom). xobs and I personally used the 840 EVO in our own units for several months, and it performed admirably.

An important metric for us was how well the drives held up under unexpected power outages. Outages happen fairly often, for example, when you're doing development work on a power management subsystem. Some hard drives failed quite reliably (how's that for an oxymoron?) after a few unexpected power-down cycles.

For the Heirloom, we used Samsung's 850 PRO series. This drive came with a serious warranty fit for an heirloom: 10 years. Samsung could offer such a high claim of reliability because the drive used a technology the company calls V-NAND, which I consider the first bona fide production-grade 3D transistor technology.

NOTE *Intel* claims *it makes 3D transistors, but that's just marketing hype. Yes, the gate region has a raised surface topology, but you still only get a single layer of devices. From a design standpoint, you're still working with a 2D graph of devices. Intel should have stuck with what I consider the "original" (and more descriptive/less misleading) name, FinFET, because by calling these 3D transistors, I don't know what it will call* actual *3D arrays of transistors, if it ever gets around to making them.*

Chipworks, a patent support company, did an excellent initial analysis of V-NAND,* showing that the technology isn't about stacking just a couple of transistors. A V-NAND stack is a 38-layer active transistor sandwich, all in a single spot. This is process technology badassery at its finest. This is Neo decoding the Matrix. This is Mal shooting first. It's a game changer, and it's not vaporware. Heirloom backers received laptops with over 4 trillion of those transistors packed inside.

Finalizing Firmware

From the software side, the next step at this point was finalizing the kernel, bootloader, and distro selection, as well as deciding what to show when Novena booted for the first time.

Marek Vasut got Novena supported in mainline U-Boot (Universal Bootloader), one of the most popular open source bootloaders. (Marek is one of U-Boot's maintainers.) The process involved a surprising number of patches, in part because few ARM boards support as much RAM as Novena. With those patches in place, Novena had full U-Boot support, including USB and video.

We decided to make Debian the factory-default distribution for Novena, and we used the stock Linux kernel with those patches added. Any patches that we thought might be useful to other projects were submitted upstream and will continue to be submitted. *Upstreaming* just means that a package that is part of a derivative operating system becomes part of the distro it's derived from.

We did keep a few local patches, ranging from specialized hacks to experimental features, features that weren't ready to push upstream, or features that relied on features that weren't upstream at the time. For example, the display system on a laptop is very different from what you'd usually

* If you're curious, you can find that analysis at *https://www.chipworks.com/about-chipworks/overview/blog/second-shoe-drops-%E2%80%93-samsung-v-nand-flash/*.

see on an ARM device. In most ARM devices, the screen is fixed during boot and it isn't possible to hot-swap displays at runtime. Like a typical laptop, Novena supports two different displays at once and allows you to plug in an HDMI monitor without requiring a reboot. Support for this feature required a local-only patch to the kernel, as it relied on features that weren't yet upstreamed for the ARM platform at that time.

Finally, we just had to decide what to show when Novena powered up. In Linux, it's not at all common to have a first-boot setup screen where you create your user, set the time, and configure the network. That's common in Windows and OS X, which come preinstalled, but under Linux, the installer generally takes care of that.

We were torn between creating a good desktop-style experience and making a practical embedded developer's experience. A desktop-style experience would ship as a blank slate and prompt the user to create an account via a locally attached keyboard and monitor. But embedded developers may never plug in a monitor, and instead prefer to connect via console or SSH; for them, a default username, password, and hostname would have been more helpful. Either way, we wanted to create just a single firmware common across all platforms and avoid special-casing releases to a particular target.

In the end, we decided to create a desktop-style experience, with escapes for power users to bypass the formalities of user enrollment. This gave us the best of both worlds. It improved the accessibility of Novena to entry-level users, yet power users could still cut to the chase and get down to work.

BUILDING A COMMUNITY

From the start, xobs and I built Novena to empower hackers, so I was pleased that even before shipping, Novena had active alpha developers. Jon Nettleton and Russell King worked on graphics, Marek Vasut from U-Boot lent a hand, and a couple of other alpha user groups actually made hardware for the system.

MyriadRF, an open source hardware and software community focused on wireless technology, created a software-defined radio board for Novena. We bought and integrated those boards with the first desktop and laptop units we shipped.

The CrypTech group also started applying Novena to its projects before the laptop shipped. The CrypTech project developed a hardware security module, with a BSD and CC BY-SA 3.0 licensed reference design. The group wanted to create a widely reviewed, designed-for-crypto device that anyone could compose for their application and easily build with their own trusted supply chain. CrypTech used Novena to prototype elements of its design.

A prototype CrypTech expansion board, plugged into the Novena motherboard

The expansion board shown here is a prototype noise source based on avalanche noise from a transistor in the middle of the board. CrypTech uses that noise to generate entropy in

Novena's FPGA. The entropy is then combined with entropy generated by ring oscillators in the FPGA and mixed using, say, SHA-512 to generate seeds. The seeds are then used to initialize the ChaCha stream cipher, ultimately resulting in a stream of cryptographically sound random values. The result is a high-performance, state-of-the art, random-number-generator coprocessor.

CLOSING THOUGHTS

As a final note, if there's one thing xobs and I have learned in the hardware business, it's that you can't count your chickens before they hatch. Making good progress to a certain point didn't mean we'd have an easy path to finished units. Even though we had fully functional prototypes at the close of fundraising, it still took months of intense effort to deliver hundreds of units to end users.

Now that Novena has finished shipping, we're continuing to support our enthusiastic yet very patient user base. It's a lot of work, which falls primarily on xobs's shoulders, but we've been answering questions from users, pushing patches, and keeping the Novena kernel up to date.

We do this even though we garner no new revenue from Novena sales. Upon reviewing our post-campaign sales data, it was fairly clear there was no viable path forward to run a hardware business selling Novena; we'd sell on average a couple of units per month. Although we cleared the minimum-order requirements of our vendors through the initial crowd-funding campaign, it would be very difficult to engage any of our suppliers at volumes less than a couple hundred units. Selling a couple units per month at that minimum buy would leave us saddled with inventory debt for about a hundred months. We'd be in debt to our suppliers for several years. Being unable to repay your suppliers for several years is also known as bankruptcy.

We are, of course, keeping our original promise to support the Novena motherboard for at least five years from the initial funding campaign. We've set aside a hefty chunk of cash to ensure a steady supply of the mainboards. Our original crowd funding and now online sales partner, Crowd Supply, has taken over the remaining inventory of cases and accessories. Thanks to our open hardware model, Crowd Supply has the option to manufacture and sell accessories for Novena, should end user demand materialize.

In the end, I'm very happy to see the tender green shoots of new projects aiming to offer better open source laptop solutions to end users. Rather than compete with them, I think it's most appropriate for Novena to give way and enable enthusiastic new developers to find opportunity and fortune selling their solutions. After all, we started on this adventure mostly to see if it could be done. We wanted to build a cool tool, customized for our everyday use case; we didn't want to start a business selling laptops with a sustainable mass-market appeal. If the ultimate impact of the Novena project is raising the bar for open hardware, and perhaps even encouraging a new generation of laptop-themed projects, that would be a huge reward in and of itself.

8. chibitronics: creating circuit stickers

In today's world of contract manufacturing and turnkey service providers, designers tend to pick from a palette of existing processes to develop products. Most consumer electronic devices are an amalgamation of rigid PCBs with SMT reflow or through-hole wave soldering, ABS or PC injection molding, sheet-metal forming, and some finishing processes like painting or electroplating. These options cover the full range of utility most products require. Really outstanding products, however, also tend to introduce new materials or novel manufacturing processes.

Developing those new processes doesn't have to be expensive—as long as you're willing to go onto the factory floor and direct the improvements yourself. In other words, the expensive bit of process development is typically paying the experts developing and qualifying the process, not so much the equipment or materials.

To prove that point to myself, I started exploring flex circuits as a design medium. Instead of using a 1- or 2-millimeter-thick rigid substrate composed of woven glass fiber impregnated with a stiff epoxy, flex circuits typically use a pliable polymer substrate just fractions of a millimeter thick. Polyimide is a popular substrate in flex circuits because of its ability to withstand soldering temperatures. Although flex-circuit technology is common inside consumer products (a mobile phone probably contains at least a half-dozen flex PCBs, connecting peripherals like buttons, cameras, and displays to the mainboard), this technology is underrepresented in hobby and DIY products. But I don't think it has to be.

I had a hunch that the right kind of product designed in flex could enable new and creative applications, but I wasn't quite sure how, so I decided to learn more about the unique benefits and challenges of designing for flexible circuits. As part of a project where I explored the guts of SD cards, which I'll talk more about in Chapter 9, I needed to create an adapter for my Novena that would allow me to snoop and emulate the NAND flash memory found inside certain styles of older SD cards. The thinness and pliability of flexible circuits were a great match for the job.

The resulting adapter was very thin; it fit perfectly under the TSOP package of the NAND. The bendy nature of the board meant I could also accommodate a broad variety of target board shapes, even boards much larger than a typical SD card. Although a useful application of flexible circuits, it still felt like I was just scratching the surface of possibility.

My custom flex adapter

Then came the moment of serendipity. While working on the SD card project, I met Jie Qi, then a PhD candidate at the MIT Media Lab, who was combining papercraft and electronics as part of her research. She was part of the group of MIT Media Lab students I took on a tour of Shenzhen in January 2012, and seeing examples of her paper circuits set the gears turning in my head.

The final artwork for Jie Qi's paper circuit art piece, Pu Gong Ying Tu

A close-up of the flowers

Peeling back the painting to reveal circuitry

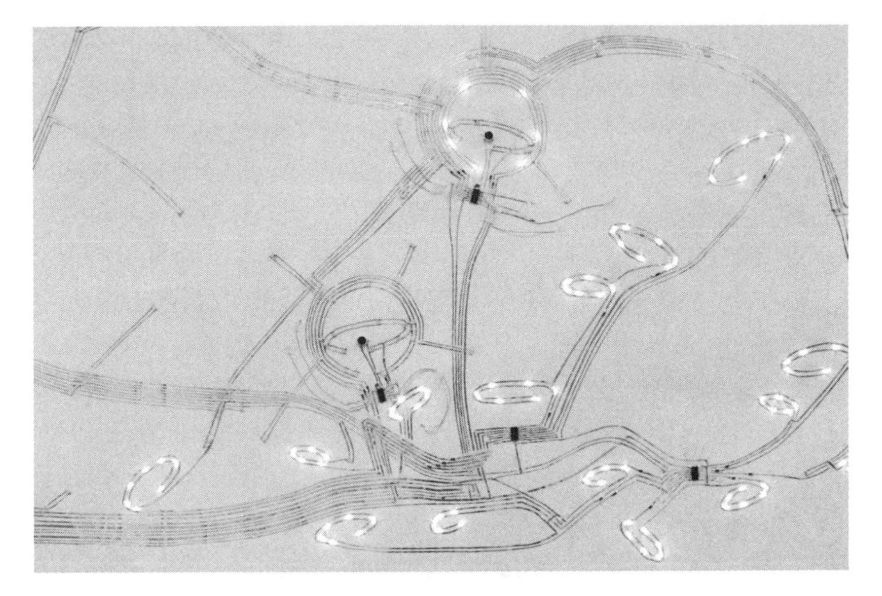

The flower circuits inside Pu Gong Ying Tu

Using nothing more than copper tape, paper, and dollops of solder or tape to hold components in place, Jie was able to craft sublime works of art that glowed and interacted with viewers. These enchanting masterpieces showed how electronics could be used not just as a functional medium, but also as an expressive medium, inspiring wonder and awe. The photo here shows the insides of one of her famous early works, *Pu Gong Ying Tu (Dandelion Painting)*, where the circuitry itself is as much a work of art as the painting overlaying it.

Jie is also very passionate about education, and she saw great potential in paper electronics to make technology more relevant and accessible to non-engineering audiences. On our trip to Shenzhen, we discussed the possibility of building circuits on flex and then soldering a flex circuit onto paper. In the end, she felt that would be at best a marginal improvement. Although soldering isn't a difficult skill to master, the high temperatures, chemicals, and specialized equipment involved are a major deterrent to beginners. What would really be magical is if circuits could be assembled like stickers on a page. Wouldn't it be great if we could use flex-circuit technology with traditional SMT reflow processes to create modules that users could then stick onto wires made of copper tape?

And that's how we came to collaborate on Chibitronics, a project in which we designed a set of peel-and-stick electronic circuits for crafting and education. Chibitronics has been an open hardware project from the start, and you can still find all the activities from the *Circuit Sticker Sketchbook*, the source code for all microcontrollers used, and other technical details through the project's wiki at *http://chibitronics.com/wiki/*.

The Chibitronics STEM Starter Kit includes the Circuit Sticker Sketchbook*, LED stickers, copper tape, batteries, and binder clips for the batteries.*

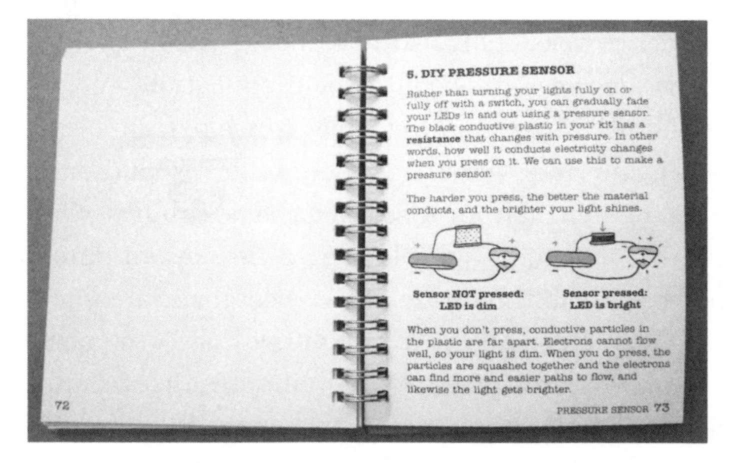

An explanation of how to create a DIY pressure sensor

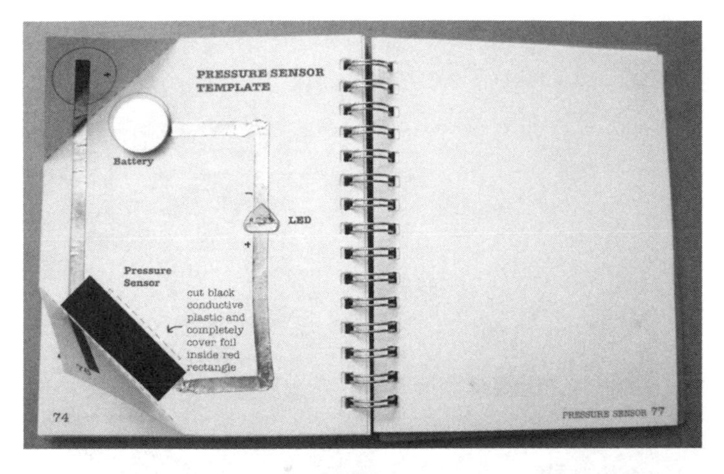

The crafted DIY pressure sensor

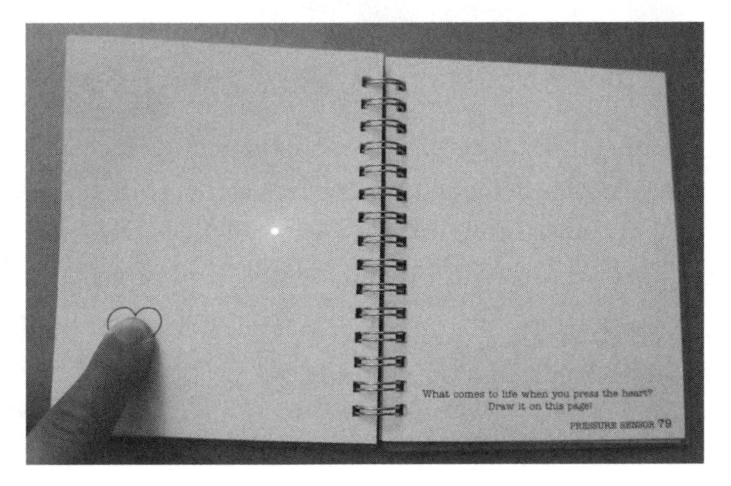

The DIY pressure sensor with paper overlay

CRAFTING WITH CIRCUITS

The solution we arrived at in early 2012 built on a body of work from Professor Leah Buechley's High-Low Tech research group at MIT. We decided to build circuits on a flexible polyimide substrate with *anisotropic tape* (also called *Z-tape*, because electricity only flows vertically through the tape, not laterally) laminated on the back.

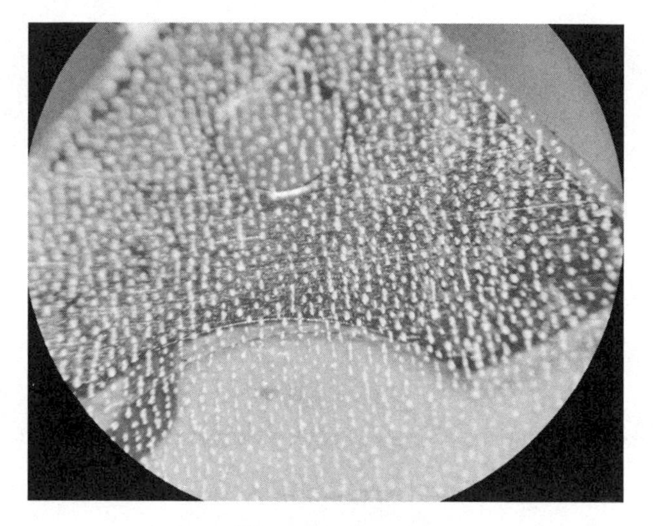

A piece of Z-tape under a microscope

Using Z-tape allows end users to assemble circuits without high-temperature processes like soldering or reflow. The ability to simply stick components in place is incredibly useful for art projects, which often involve heat-sensitive and/or pliable material substrates like paper, fabric, and plastic. Circuit stickers and copper tape are flexible, too, further enabling anyone to integrate electronics into projects using nontraditional materials. Such friendly and expressive materials encourage creators to turn the circuits themselves into beautiful works of art.

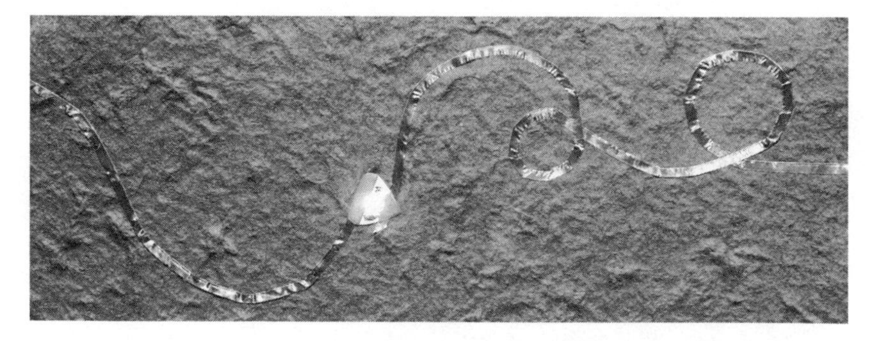

Circuit stickers on paper

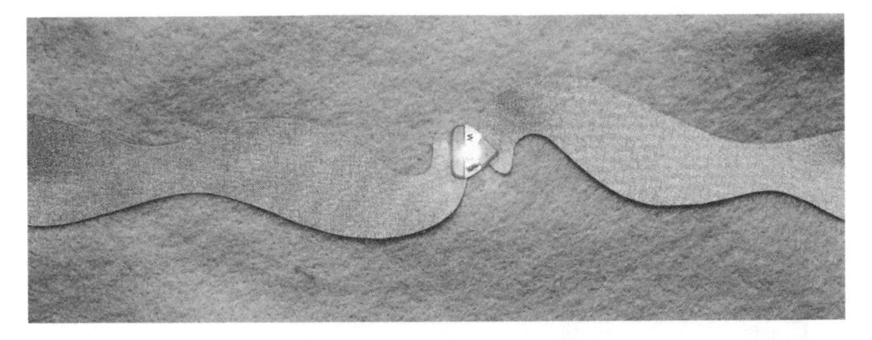

Circuit stickers on fabric

Creating these circuit stickers revolved around the limitations of the Z-tape. In the magnified section of Z-tape laminated onto a polyimide substrate shown here, the silvery-white stipples are tiny metal particles that span from one side of the adhesive layer to the other according to a statistical distribution. Given the nature of the metal distribution, to ensure good electrical contact, each pad on a circuit sticker needed to be fairly large. Furthermore, traces very close to each other could be shorted out by the embedded metal particles, so as I designed the circuits, I had to be careful to leave enough space between exposed pads. The datasheet for the Z-tape material contains rules for the minimum pad size and spacing, so I used those as a guide.

Developing a New Process

It's one thing to design stickers containing working electronic circuits, but it's a whole different thing to actually build them. No standard manufacturing processes existed that could produce circuit stickers as we envisioned them. At last, I had a meaningful opportunity to test my theory that new process development can be done cheaply if you're willing to do it yourself. So I started my own little research program to explore flex-circuit media and the challenges of making circuit stickers out of them, all on a shoestring R&D budget.

Visiting the Factory

As a first step, I visited the facility where flex PCBs are manufactured. The visit was eye-opening.

A worker manually aligning coverlay onto flex-circuit material

Instead of soldermask, flex-circuit traces are protected by a polyimide sheet called *coverlay*. Soldermask is too brittle and will crack if bent, but coverlay reliably stays intact over thousands of flexing cycles. Sometimes, however, you want to make portions of a flex circuit stiff; for instance, a part of the circuit might need to stay stiff for mechanical mounting, and a stiff circuit is also helpful for SMT processing.

Steel plates being laminated to the back of flex-circuit material

I knew that polyimide stiffeners could be laminated to flex, but as it turns out, steel lamination is also possible. I wouldn't have known that if I hadn't taken the factory tour myself. Visiting the factory in person also gave me an invaluable opportunity to see the wide range of complex shapes that could be produced thanks to die cutting. Having a variety of possible shapes was key, because we wanted to make the circuit stickers look cool, too. Questions like how narrow we could cut the material or how tight a radius is allowable in a die cut are difficult to answer by email, but the answers were intuitively obvious after I saw the process in person.

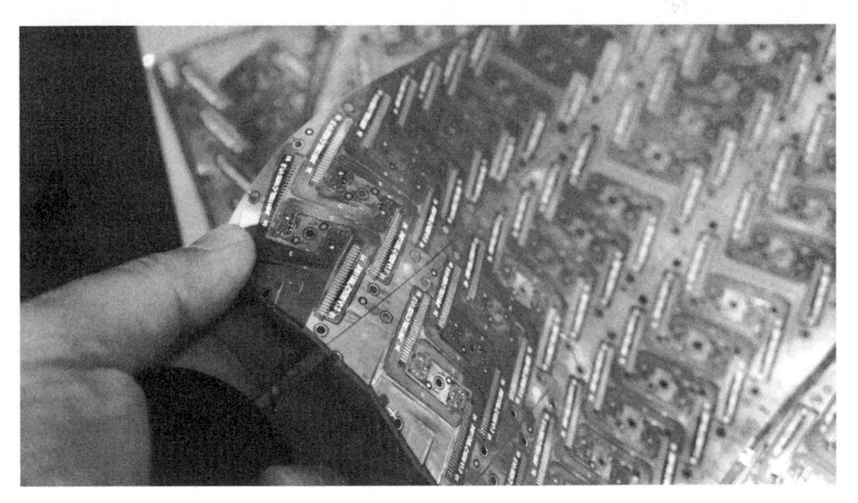

The intricate flex-circuit shapes achievable with die cutting

Performing a Process Capability Test

After the factory visit, the next step was to do a *process capability test* to push the limits of the manufacturing process. We designed a non-homogenous sheet of sticker variants that exercised all kinds of capabilities: long via chains, 3-mil line widths, 0201 components (a small SMT package size), 0.5 mm pitch QFN parts (surface-mount components that have all their contacts on the bottom), bulky components, the use of soldermask instead of coverlay, fine detail in silkscreening, captive tabs, curved cutouts, hybrid SMT and through-hole soldering techniques, Z-tape lamination, and more. Our process capability test intentionally broke parts of the manufacturing process to discover weak links that could prevent our design from working out.

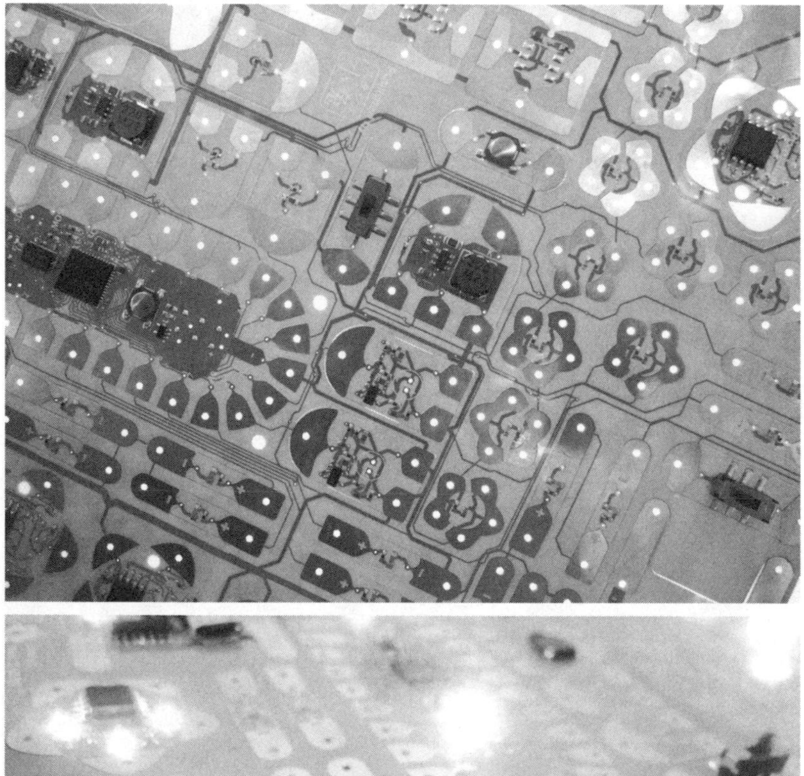

The circuit sticker design we manufactured for the process capability test

When I first presented the design, the factory rejected it outright, saying it was impossible to manufacture. After I explained my goals, however, the factory agreed to produce it, with the understanding that I'd accept and pay for all units made, naturally including the defective ones. Through analyzing the failure modes of the defective units, I developed a set of design rules for maintaining high yield (and therefore lowering cost) on the circuit stickers.

Based on these design rules, Jie and I created our first set of "production candidate" stickers. They included LEDs of four different colors (white, red, blue, and yellow), as well as two sets of smart stickers. The first set of smart stickers contained a preprogrammed microcontroller that could generate patterns of light, such as fading, heartbeats, twinkling, and blinking. We called these the "effects" stickers; they are a form of *physical programming* that enables noncoders to customize the behavior of their projects. The second set contained a user-programmable microcontroller with a fun record-and-playback capability loaded into it as a demo, along with three sensors. We called these the "sensor & microcontroller" stickers.

We ran small batches of our production candidates to find problems we might encounter should we need to scale up, and we thoroughly investigated any issues that would affect reliability, yield, or usability. In particular, we had to develop a novel method for laminating Z-tape onto the back of the stickers that would be process-compatible with the type of die cutting necessary to create stickers.

After two iterations of production candidates, we felt we were ready to see what other people could do with circuit stickers. As this was part of Jie's doctoral research, we had two options for doing user testing. The traditional academic approach would have been to apply for a budget from her advisor, produce a limited number of stickers, and conduct a series of closed workshops to study how young and creative minds interacted with this new media. But this happened in 2013, so viable crowdfunding platforms unlocked the possibility of offering our research directly to interested users, thus allowing us to conduct research at scale. The MIT Media Lab where Jie researched is also very keyed in to the possibilities enabled by research at scale, as embodied by their "deploy" initiative. In 2011, when Joi Ito became the Media Lab's new director, he started transforming the Media Lab's culture from "demo or die" to "deploy or die," which was eventually shortened to the less menacing "deploy" directive. Under the old "demo or die" regime, research groups were encouraged to create whizzy demonstrations of technology that could help raise money. Under Ito's directive, the idea is to get technology out of the lab and into the wild by conducting research at scale through tools like crowdfunding and lean hardware.

In November 2013, we launched a crowdfunding campaign with Crowd Supply. It was very important to us to remain pure to the academic mission behind the circuit stickers, so we set our funding goal at just $1. If even one person thought circuit stickers might be interesting, we'd produce the stickers and work with that person to gather feedback. And, of course, we would make that research available to the world, in case someone wanted to fork the project or otherwise hack their circuit stickers.

We beat our modest goal by several orders of magnitude, closing just shy of $60,000 after a little over one month of funding and a very low-key campaign.

DELIVERING ON A PROMISE

As part of our campaign, we stated that we would ship orders for fulfillment by May 2014. Thankfully, we were able to meet our goal, right on time.

Sixty-two cartons containing over a thousand Chibitronics starter kits, waiting for pickup

Delivering on time is no simple task for any crowdfunded project, however. I made the contentious choice to use Crowd Supply in part because they show more savvy around vetting hardware products, and the services they offer to campaigns (fulfillment, tier-one customer support, post-campaign preorder support, and rolling delivery dates based on demand versus capacity) are a boon for hardware upstarts. Getting fulfillment, customer support, and an ongoing e-commerce site as part of the package meant we didn't have to hire someone to deal with all of that. Whether your "company" consists of just two people trialing an academic project or a couple of people working out of a garage, that's a big deal.

Crowd Supply doesn't have the same media footprint or brand power that Kickstarter has, which can make it harder to raise as much money. But at the end of the day, I feel it's very important to establish an example of sustainable crowdfunding practices that's better for both the entrepreneur and the consumer. It's not just about a money grab today; it's about building a brand and reputation that can be trusted for years to come.

WHY ON-TIME DELIVERY IS IMPORTANT

I set a personal challenge for Chibitronics to take our delivery commitment to backers very seriously. I've seen too many underperforming crowdfunding campaigns, and I'm deeply concerned that crowdfunding for hardware is becoming synonymous with scams and spams.

Kickstarter and Indiegogo have been plagued by nondelivery and scams, and their blithe, *caveat emptor* attitude around campaigns highlights the conflict of interest between consumers and crowdfunding websites. The crowdfunding sites are basically saying to backers, "Hey, thanks for the nickel, but what happened to your dollar is your problem." I'm honestly worried that crowdfunding will get such a bad reputation that it eventually won't be a viable platform for well-intentioned entrepreneurs and innovators.

The bottom line is this: if I can't prove to current and future backers that I can deliver a project on time, I stand to lose a valuable platform for launching my future products. Fortunately, we definitely proved ourselves with Chibitronics, and I've continued to use Crowd Supply for other crowdfunding projects since.

LESSONS LEARNED

We didn't deliver Chibitronics on time because we had it easy, though. When I drew up the original campaign timeline, my minimum and maximum bounds on delivery time spanned from just after Chinese New Year 2014 (February) to around April. I padded that schedule by one month beyond the max, just to be safe, and we used every last bit of this padding.

I made a lot of mistakes along the way, but through a combination of hard work, luck, planning, and strong factory relationships, we successfully overcame many hardships. Here are a few lessons I learned during the process.

Not All Simple Requests Are Simple for Everyone

Every Chibitronics starter kit included a physical copy of a fantastic book Jie wrote as a step-by-step, self-instruction guide to designing with circuit stickers, the *Circuit Sticker Sketchbook* (shown on pages 256–257). The book is unusual because you're meant to paste electronic circuits into it, so we had to customize several aspects of the printing. The paper had to be the right thickness to get good light diffusion when LEDs were placed underneath a sheet. The binding needed special attention for a better circuit-crafting experience, and there's even a little pocket in the back to hold swatches of craft material used as part of the projects in the book.

The printer found most of these requests relatively easy to accommodate, but one in particular threw them for a loop. The book's metal spiral binding had to be nonconductive so that placing copper tape on the binding wouldn't accidentally cause a short circuit.

Checking a wire for conductivity seems like a simple enough request for someone who designs circuits for a living, but for a book printer, it's weird. No part of traditional book printing or binding requires such knowledge. The printer originally said they couldn't guarantee anything about the conductivity of the binding wire. Sure enough, while the first sample wire was nonconductive, the second was conductive, and the printer couldn't explain why.

Face-to-face meetings were invaluable here. Instead of yelling at the printer over email, we arranged a meeting with them during one of my monthly trips to Shenzhen. We had a productive discussion about their concerns, and at the conclusion of the meeting, we ordered them a $5 multimeter in exchange for a guarantee of a nonconductive book spine. In the end, the printer was simply unwilling to guarantee something for which they had no quality control procedure, which is completely reasonable. We just had to teach them how to use a multimeter.

This unusual nonconductivity requirement did extend our lead time by several days and added a few cents to the cost of the book, but overall, I was willing to accept that compromise.

Never Skip a Check Plot

The pad shapes for the circuit stickers are complex polyline geometries, which Altium, the PCB design software I was using, didn't handle very gracefully. I discovered the hard way that in Altium, the soldermask layer occasionally disappears for pads with complex geometry. Older versions of my design would contain a soldermask layer, but then upon saving the design file, the layer would silently disappear. This sort of bug is rare, but it does happen. Normally, I'd import the gerber file into a third-party tool as a check plot before making an order, but I was in a rush and reordering an existing design that had worked before, so I skipped the check plot procedure.

The result? Thousands of dollars' worth of PCBs had to be scrapped, and we lost four weeks from the schedule. Ouch.

It was good that I padded my delivery dates—and that I keep a bottle of fine Scotch on hand, to help bitter reminders of what happens when I get complacent go down a little easier.

If a Component Can Be Placed Incorrectly, It Will Be

I'm paranoid about parts being placed incorrectly, as this problem has burned me many times. The Chibitronics effects sticker sheet was a prime example of the issue waiting to happen.

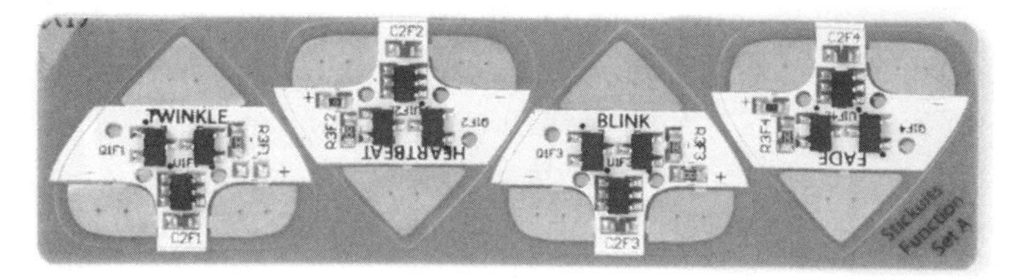

The Chibitronics effects stickers

The sheet is an array of four stickers that flash different patterns on an LED but are otherwise identical. The flashing pattern is controlled by software. Trying to manage four separate firmware files and get them all loaded into the right spot in a tester is a nightmare waiting to happen. To solve that problem, I designed the stickers to use the exact same firmware. Their behaviors were instead set by the value of a single external resistor, which was measured on boot by the microcontroller's integrated ADC.

My logic went something like this: if all the stickers have the same firmware, there's no "wrong way" to program the stickers. Right?

Unfortunately, I also designed the master PCB panels to be perfectly symmetric. You could load the panels into the assembly robot rotated by pi radians, and the assembly program would run flawlessly—except that the resistors setting the firmware behavior would be populated in reverse order compared to the silkscreen labels. Despite having fiducial holes to provide a frame of reference and text on the PCBs in both Chinese and English that is uniquely orienting, this problem actually happened. On the first effect sticker samples, the "heartbeat" sticker was "blinking," the "twinkle" sticker was "fading," and vice versa.

Fortunately, the factory very consistently loaded the boards in backward, which is the best case for a problem like this. I rushed a firmware patch (also a risky thing to do) that reversed the interpretation of the resistor values, and had a new set of samples shipped to me in Singapore via FedEx for a sanity check. We also built a secondary test jig to add a manual double-check for correct flashing behavior on the line in China.

The effects sheet problem was solved, but in making that additional test, we discovered another common problem.

Some Concepts Don't Translate into Chinese Well

I wrote instructions in Chinese to describe the difference between fading (a slow blinking pattern) and twinkling (a flickering pattern) to the factory, but it turns out that the Chinese translations for *blink* and *twinkle* are similar. *Twinkle* translates to 闪烁 ("flickering, twinkling") or 闪耀 ("to glint, to glitter, to sparkle"), and *blink* translates to 闪闪 ("flickering, sparkling, glittering") or 闪亮 ("brilliant, shiny, to glisten, to twinkle").

I always dread writing subjective descriptions for test operators in Chinese, which is part of the reason I try to automate as many tests as possible. As one of my Chinese friends once remarked, Mandarin is a wonderful language for poetry and arts but difficult for precise technical communications.

The challenge, then, was to come up with a bulletproof, cross-cultural explanation of the difference between fading and twinkling, using only simple terms anyone could understand; that is, I had to avoid technical terms like *random*, *frequency*, *hertz*, and *periodic*.

I sent the factory a video of the different LED patterns, and our factory recommended we use 渐变 ("gradual change") for *fade* and 闪烁 ("flickering, twinkling") for *twinkle*. I'm still not convinced that was a bulletproof description, but it was superior to any translation I came up with. And, to this day, we are dogged by problems trying to explain to quality control staff the difference between these effects. It turns out that a malfunctioning sticker *also* makes a pretty good twinkling effect—for a while.

Funnily enough, it was also a challenge for Jie and me to agree upon what a "twinkle" effect should look like. She described our first iteration of the effect as "closer to a lightning storm than twinkling." We had several long conversations on the topic, followed by demo videos to clarify the desired effect.

We basically tweaked code until it looked about right to both of us. Given the difficulty we had describing the effect to each other, it's no surprise I had trouble accurately describing the effect in Chinese.

Eliminate Single Points of Failure

When we built test jigs, we built two copies of each, even though throughput requirements demanded just one. Why? Because one might fail.

And guess what: one test jig did fail. I still don't know why. Thank goodness we built two copies, though, or I'd have had to rush to China on short notice to diagnose why our sole test jig didn't work.

Some Last-Minute Changes Are Worth It

About six weeks before we finalized our order for the Chibitronics kits with the factory, Jie suggested that we include a stencil of the sticker patterns with the sensor and microcontroller kits. She reasoned that it can be difficult to lay out the copper tape patterns for complex stickers like the microcontroller, which has seven pads, without a drawing of the contact patterns. I originally resisted the idea; I didn't want to delay shipment on account of something we didn't originally promise. As Jie discovered, I can be very temperamental, especially when it comes to schedule slips. (Sorry, Jie! Thanks for bearing with me.)

But her arguments were sound, so I instructed our factory to search for a stencil vendor. After two weeks, we hadn't found anyone willing to take the job, but our factory's sourcing department didn't give up. Eventually, they found one vendor who had enough material in stock to tool up a die cutter and turn around a couple thousand stencils within two weeks—just barely in time to meet the schedule.

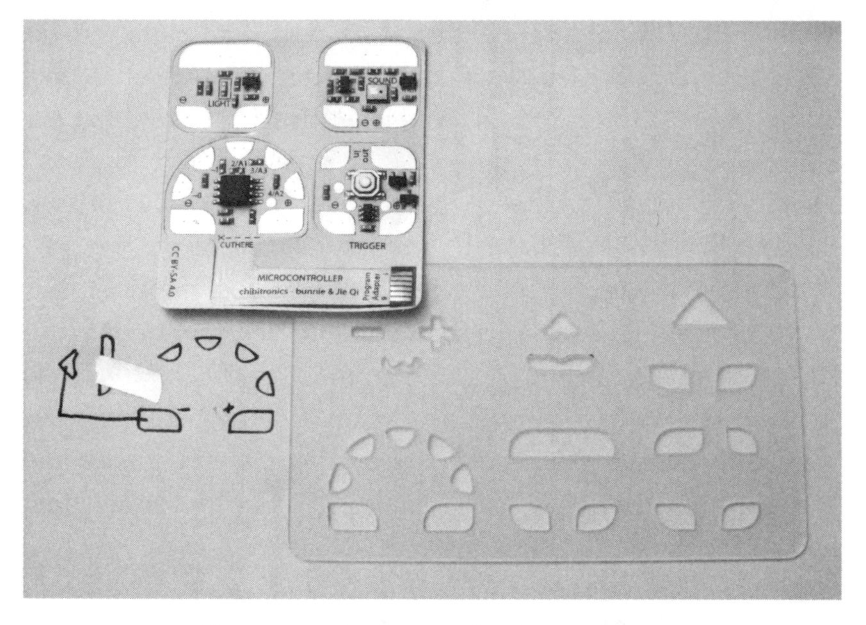

The sensor and microcontroller sheet and stencil

When I got samples of the sensor and microcontroller kit with the stencils, I gave them a whirl. Jie was absolutely right about their utility. I found my experience vastly improved when I had a template to work from, particularly for the microcontroller sticker with seven closely spaced pads, and I felt users would agree. That's how even though the stencil wasn't promised as part of the original campaign, all backers who ordered the sensor and microcontroller kit received a free stencil to help them lay out designs.

Chinese New Year Impacts the Supply Chain

Even though the Chinese New Year is a two-week holiday, our initial schedule essentially wrote off the month of February. Reality matched this expectation, but I want to share with you exactly how Chinese New Year impacted this project, in case you're considering manufacturing a product in China.

We had a draft manuscript of our book ready in January, but I couldn't get a complete sample until March. That wasn't because the printer was closed for a month straight; like everyone else, their holiday was about two weeks long. The paper vendor, however, started their holiday about 10 days before the printer, and the binding vendor ended their holiday about 10 days after the printer. Even though each vendor took only two weeks off, the net supply chain for printing a custom book was out for around 24 days, or effectively the entire month of February. The staggered observance of Chinese New Year is necessary because of the sheer magnitude of human migration that accompanies the holiday.

Shipping Is Expensive and Difficult

When I ran the initial numbers on shipping, I realized that we weren't exactly selling circuit stickers—taking the book into account, by volume and weight, our principal product was printed paper. To optimize logistics cost, I pushed to ship starter kits (which contained a book) and additional standalone book orders by ocean, rather than air.

We actually had starter kits and books ready to go almost four weeks before the first kits shipped, but we just couldn't get a reasonable quotation for the cost of shipping them by ocean. We spent almost three weeks haggling and quoting with ocean freight companies. In the end, their price was basically the same as going by air but would take three weeks longer and incurred more risk. Freight cost is apparently a minor component of shipping by ocean, and you get killed by a multitude of surcharges, from paying the longshoremen to paying all the intermediate brokers and warehouses that handle your goods at the dock. Those fixed costs added up such that even though we were shipping over 60 cartons of goods, air shipping was still more cost-effective.

NOTE *For reference, a Maersk 40-foot sea container would fit over 1,250 cartons, each containing 40 starter kits. We were an order of magnitude away from being able to efficiently utilize ocean freight.*

You're Not Out of the Woods Until You Ship

At each milestone in this project, I had to remind myself not to count my chickens before they hatched. Problems ranging from a routine UPS screwup to a tragic aviation accident to a logistics problem at Crowd Supply's fulfillment depot to a customs problem could stymie an on-time delivery. But, at the very least, we did everything within our power to deliver on time.

Thankfully, when all was said and done, our backers received their orders right on time. Since then, Chibitronics has continued to surpass my wildest expectations. Although we started this project as an academic experiment, grassroots user adoption prompted us to grow the experiment into a full-fledged company. As the circuit stickers are an open hardware project, the specs are available for savvy hackers to play with, but most users are nontechnical folks who would benefit more directly from support on basic usage. To that end, the company strives to provide users with assistance, activities, and more stickers to help them keep learning and making beautiful electronic crafts.

CLOSING THOUGHTS

Chibitronics has been an ongoing learning experience for me, as I've never had a company successfully mature like this. I'm excited to see where the company goes, but as an engineer, I also know my limitations: I'm not cut out to be a businessperson. Once the company is big enough to support its own staff in a sustainable fashion, I'm looking forward to handing over the reins, returning to my workbench, and dreaming up new open hardware inventions.

Part 4

a hacker's perspective

Engineering and reverse engineering are two sides of the same coin. The best makers know how to hack their tools, and the best hackers routinely make new tools. I might set out to design a circuit, and find myself reverse engineering a chip because the datasheet is vague, incomplete, or simply incorrect. Engineering is a creative exercise; reverse engineering is a learning exercise. When you combine them, even the toughest problems can be solved as a creative learning exercise.

I spent over a quarter-century in school, but I've learned more about electronics from reverse engineering. I love trying to figure out why the engineer behind a piece of random hardware made certain design choices. Highly skilled engineers develop clever tricks without realizing how innovative they are. Those tricks often go undocumented or unpatented, and the only way to tap that knowledge is to decipher it from finished designs.

After seeing enough boards, I started recognizing patterns and personal styles that almost have a cultural nature about them. For example, Apple circuit boards are austere and black, with a look almost as iconic as Steve Jobs's black mock turtlenecks. There are so many decisions to make when designing a circuit board that most engineers can only draw from their cultural influences and toolchains to constrain stylistic things like fonts and part choices.

This kind of learning is so important to me that, for over a decade now, every month I've presented a circuit board on my blog and challenged readers to divine its function from its design. Part of my motivation for holding these regular competitions is to make reverse engineering feel culturally acceptable to readers. People often ask me if reading other people's designs or modifying and hacking hardware is legal. But anyone who has raised a child knows that learning through emulation is a part of human nature. I disagree with interpretations of the law that put the terms of a software license above your right to own your hardware. If you can't hack it, you don't own it.

The importance of democratic access to technology only grows as we become increasingly dependent on smartphones and computers. Technology is fundamentally neutral toward human ethics; the people who control technology are responsible for applying it ethically. One school of thought believes that technology should be controlled by a select group of trusted masters; the other believes that control over technology should belong to anyone with the motivation and will to learn it. Increasingly, our technology infrastructure is becoming a monoculture managed by a cartel of technology providers. Everyone carries identical phones running operating systems based on the same libraries and uses one or two cloud services to store their data. But history has proven that a monoculture with no immunity is a recipe for disaster. One virus can wipe out a whole population. Universal access to technology may allow the occasional bad actor to develop a harmful exploit, but this bitter pill ultimately inoculates our technological immune system, forcing us to grow stronger and more resilient. Wherever that threat comes from, a robust and vibrant culture of free-thinking technologists will be our ultimate defense against any attack.

Speaking of viruses and immune systems, there are remarkable parallels between hardware systems and biological systems. Just as hacking is all about rethinking APIs to do unexpected things, a central tenant of biology—evolution—is all about superior implementations of "APIs" superseding weaker interpretations.

I routinely read journals about the life sciences not just because I find the subject fascinating, but also because it's good for me. Looking outside your primary field for fresh ideas is very helpful for problem solving. Figuring out how an organism works is an incredibly difficult reverse engineering problem: there's no documentation, there's no designer to consult, and your diagnostic tools are roughly equivalent to throwing crate after crate of smartphones into a blender and running the mixture through various sieves. Biologists have developed a bag of extremely clever tricks to map out complex systems without the benefit of an oscilloscope, and at a high level, some of the principles are applicable to electronic systems.

As our understanding of biology becomes more complete, there's ample opportunity for computer engineering principles to advance the field. We're already at the point of custom-engineering organisms; the technology to hack humans—or engineer our successor—is likely to arrive within decades. Such powerful tools deserve a closer look so that we can make independent judgments about what is fact and what is fiction.

While engineering is a creative activity, hacking is an important and often underrated learning exercise. The ability to effortlessly switch modes from forward to reverse engineering is a powerful tool, and the right to hack is the foundation of a healthy technological culture. The first chapter in this section reviews some of my own hacking methods and efforts and discusses some of the legal frameworks that protect these activities. The second chapter attempts to unpack some key concepts from biology and frame them from the perspective of an electronics person. The final chapter in this book is a collection of interviews where I discuss what being a hacker means to me, as well as recap some of my experiences in manufacturing and hardware startups. The collection isn't exhaustive, but I hope you enjoy reading some of my more off-the-cuff thoughts.

9. hardware hacking

The biggest barrier to hacking is often the fear that you'll break something while poking around. But you have to break eggs to make an omelet; likewise, you have to be willing to sacrifice devices to hack a system. Fortunately, acquiring multiple copies of a mass-produced piece of hardware is easy. I often do a bit of dumpster diving or check classified advertisements to get sample units for research purposes. I generally try to start with three copies: one to tear apart and never put back together, one to probe, and one to keep relatively pristine. I use the pristine copy to sanity-check whether a certain behavior is due to my probing or just how the hardware behaves.

My typical approach to any hardware hack is first getting the device open and then getting a probe in just the right spot without affecting the device's functionality. When you're looking inside computer chips, that's virtually the entire challenge. The first hack in this chapter is an example of silicon hacking, and you'll see that once the package is off and you're staring at naked silicon, an attacker has a profound advantage.

Some hardware hacks require more system engineering, particularly when you want to reverse engineer and repurpose a device. In these situations, I tend to develop additional bespoke tools that allow me to tweak and observe a system in close to real time, or at least as fast as I can type commands, to minimize the time spent validating hypotheses. The goal is to make the primary limitation how fast you can think of ideas to test, not how long it takes to upload a change to test those ideas. The second hack in this chapter talks about reverse engineering a relatively simple System-on-Chip (SoC) device found inside common SD memory cards and some tools I developed to aid that process.

Finally, some hacks inevitably push the boundaries of the law. The third hack in this chapter talks about NeTV, a system I developed that takes a new look at the High-Definition Content Protection (HDCP) encryption standard, which secures most HDMI video links. NeTV is a hack on both a legal issue and a hardware system. It works around the thorny problems presented by the DMCA by reinterpreting the HDCP standard to enable a man-in-the-middle (MITM) attack to change video data without circumventing encryption. No circumvention, no DMCA problem. Hacks often push the boundary of what's legal and what's been tested in the courts. Just like any other system, the legal system can also be hacked, and one key takeaway from this chapter is how to think of laws as just another constraint to work with on the way to achieving a particular goal.

The final hack in this chapter combines hardware penetration, tool creation, and legal considerations to reverse engineer a complex mobile phone SoC. That's another project I worked on with xobs, and once again, building bespoke hacking tools was invaluable because it allowed us to experiment with the system as it ran.

HACKING THE PIC18F1320

Keeping a secret is a common challenge for any security system. To solve this challenge, security system designers frequently hide secrets inside silicon chips because the chips' rugged epoxy packages and tiny geometries are difficult to penetrate and inspect.

This sounds good in theory but is problematic in practice. Chip designers make mistakes, and when a chip has a problem, the designers need a way to open it up and investigate. This situation is so common that there are commercial services that specialize in opening up chips expressly for that purpose. Called *failure analysis services*, they've mastered several techniques for removing tough epoxy from chips.

A couple of years before my crash course in setting up a Chinese supply chain with Chumby, I decided it would be fun to demonstrate how simple hacking a chip can be if you're aware of failure analysis services. At the time, Microchip's PIC series of microcontrollers was quite ubiquitous, so I decided to have a go at a popular PIC model. PICs typically have *configuration fuses*, which you can activate to prevent certain regions of memory from being read or written to. But there's often a legitimate need to read the contents of a secured, programmed PIC. For instance, a company that loses either the documentation for a product or the personnel that originally created the codes for a secured PIC would be stuck without a way to read the chip. This is a problem when a company needs to revise or upgrade a legacy line of products.

I wanted to figure out how to dump the memory from a secured PIC. Knowing I'd have to break a few eggs to make this omelet, I scored four PIC18F1320s from a friend and started stripping them down. Here's what I found.

A PIC18F1320 in its native state

Decapping the IC

First, I had to take the top off so I could see the silicon under the hood. Many homebrew techniques for decapping a chip typically involve applying fuming nitric or sulfuric acid, but those aren't compounds you'd want to keep at home, nor are they easy to obtain. Nitric acid, in particular, is an important compound for explosives fabrication. So, I've found the easiest and most reliable way to decap a chip is to just send it to a failure analysis lab. For about $50, you can have a decapped part in two days.

I decapped three parts for this project. Two were *functionally decapped* (silicon revealed with the device still in its lead frame, fully functional), and the last was *fully decapped* (just a bare silicon die with no package). I had one die fully decapped because my inspection microscope had a very short working distance at the highest magnifications, and the remaining epoxy from the package would have interfered with the lens.

A functionally decapped PIC18F1320.
The little raised square in the middle (it's goldish in real life) is the silicon chip.

Taking a Closer Look

With my decapped ICs in hand, I did a sweep around one of the dies with the microscope and noticed several prominent features. Because physics is the same everywhere, most of the fine-grained structure in a silicon chip looks pretty much the same, no matter who makes the chip. These constraints propagate their way up to the system level, and with a bit of training, you can read a silicon chip like a book.

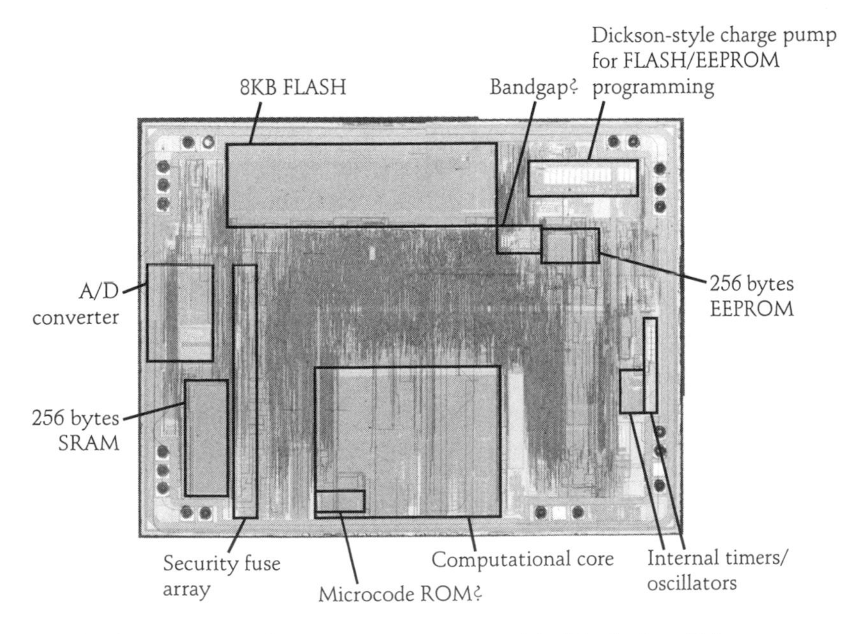

My best guess at what various structures in this chip do. I could be wrong.

One set of structures grabbed my attention immediately: there were metal shields over some transistors, following a regular pattern that had about the right number of devices to account for all the security bits. Full-metal shields covering a device are very rare in silicon, so they're like a big X marking the spot where something very important is kept.

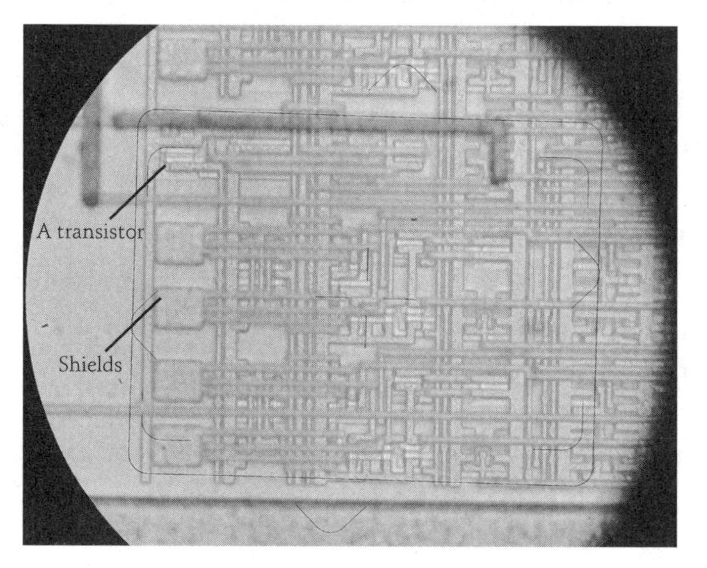

Zooming in on the metal shields

Erasing the Flash Memory

The shields were significant because of some interesting facts about flash memory technology, which this PIC device used to store the security fuse information, as well as the internal program code. Flash technology uses a floating-gate transistor structure very similar to old *UV-erasable programmable read-only memory (UV-EPROM)* technologies like the ceramic-packaged 2716 chips from the 1970s, which had quartz windows so they could be erased.

In both flash and UV-EPROM devices, data is written when electrons tunnel into a floating gate, where the electrons remain for decades. The extra electrons in the floating gate create a measurable offset in the characteristics of the storage transistor. The difference is that flash memory can withdraw the stored electrons (erase the device) using only electrical pulses, while a UV-EPROM requires energetic photons to knock the electrons out of the floating gate. The UV light required to accomplish this is typically on a wavelength of around 250 nm. You need expensive quartz optics to manipulate this wavelength of UV without excessive loss, making it a bit difficult to harness.

Here's the important conclusion I drew from these facts: flash devices can usually *also* be erased using UV light since they have a similar transistor structure to UV-EPROM devices. The encapsulation around a flash device normally prevents any UV light from effectively reaching the die, but since the PIC devices had the plastic around them removed, I could attempt to apply UV light and see what happened.

I performed a simple experiment by programming the PIC device with a ramping pattern, where I stored the hexadecimal numbers from 0x00 to 0xFF over and over again. Then, I tossed the PIC into my UV-EPROM eraser to bake for . . . oh, about the length of a good long shower and some email checking. When I took the device out of the eraser, the flash memory was indeed blanked to its normal all 1s state, and the security fuses were unaffected. After baking a few more PIC devices in the eraser, I found that if I didn't bake a PIC long enough, I got odd readings out of the array I wrote to, such as all 0s, a phenomenon that I still don't understand.

Erasing the Security Bits

Clearly, the metal shields over the security fuses were there to thwart attempts to selectively erase the security fuses while leaving the flash memory array unaffected.

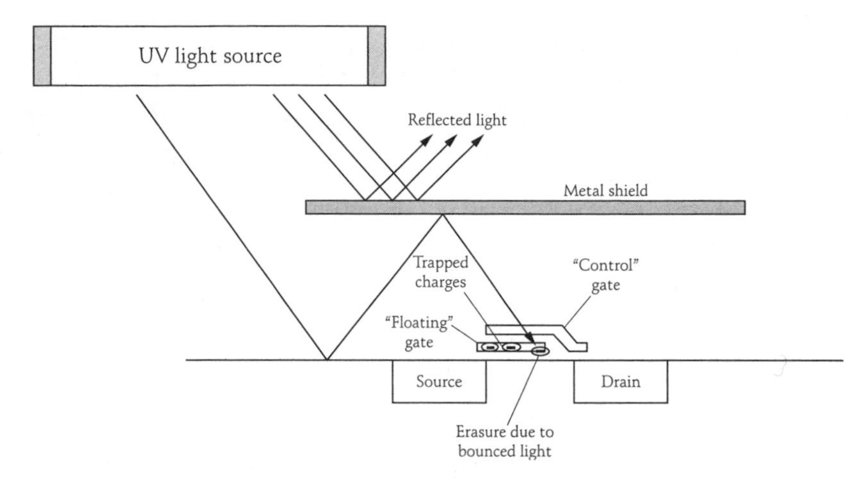

A diagram showing how the shields got in the way of the fuse bits, and how to work around them

My problem was that for the flash memory transistor to be erased, high-intensity UV light needed to strike the floating gate. The metal shield effectively reflected all incident light, so the light never reached the gate. But I knew there was a refraction index mismatch between the optically clear protective dielectric layer of silicon dioxide covering the chip and the silicon proper, meaning light at certain angles would reflect off of the smooth silicon surface. For an example of this reflective effect, jump in a swimming pool, go under water, and look up at where the water and air meet. The water should look highly reflective at an oblique angle because the refractive index mismatch between water and air causes total internal reflection of light.

I planned to use this reflection to bounce the UV light off the oxide to hit the metal shield and bounce back onto the floating gate. By angling the PIC inside the ROM eraser, I thought I could get enough light to bounce into the flash memory transistor region and erase the security bits. After a couple of attempts using bits and bobs of material to fix the angle of the chip, I developed a simple technique that worked surprisingly well: shoving the chip into the antistatic foam liner of the UV eraser at an angle.

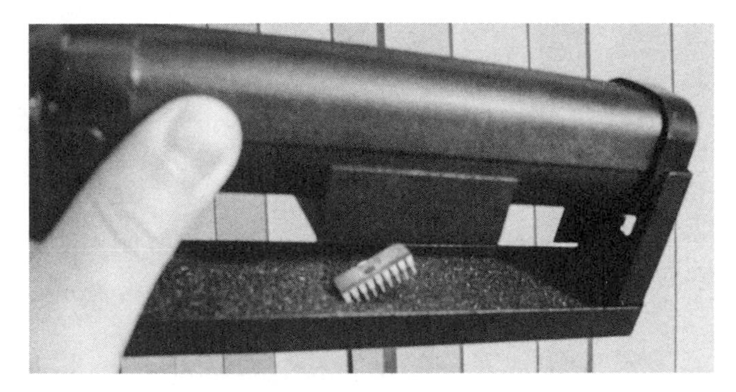

The chip in the UV eraser's antistatic foam

Protecting the Other Data

That technique didn't protect the flash data I wanted to keep, though. To avoid erasing this data, I made a hard mask out of a very carefully cut piece of electrical tape and stuck that mask to the surface of the die using a steady hand, two tweezers, and a microscope. The electrical tape blocked the UV light from directly hitting the flash code memory regions and somewhat absorbed light bounced back from the silicon substrate.

The die in its package, with electrical tape over the flash ROM array

This mask allowed me to reset only the security fuses without impacting the flash code array too much. The following screenshots show the array memory status according to the programming and readback tool I was using.

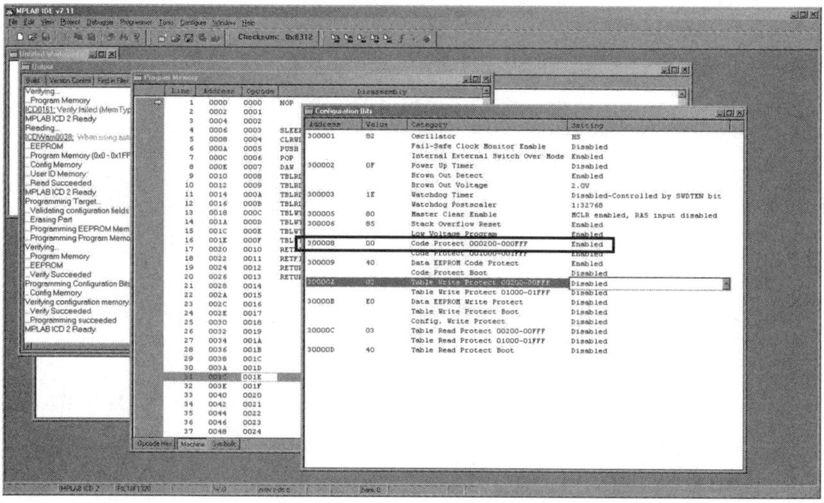

MPLAB IDE v7.11

File Edit View Project Debugger Programmer Tools Configure Window Help

Checksum: 0x8312

Output

Build | Version Control | Find in Files

Verifying...
...Program Memory
ICD0161: Verify failed (MemTyp
MPLAB ICD 2 Ready
Reading...
ICDWarn0038: When using aut
...EEPROM
...Program Memory (0x0 - 0x1FF
...Config Memory
...User ID Memory
...Read Succeeded
MPLAB ICD 2 Ready
Programming Target...
...Validating configuration fields
...Erasing Part
...Programming EEPROM Mem
...Programming Program Memo
Verifying...
...Program Memory
...EEPROM
...Verify Succeeded
Programming Configuration Bits
...Config Memory
Verifying configuration memory
...Verify Succeeded
...Programming succeeded
MPLAB ICD 2 Ready

Program Memory

Line	Address	Opcode	Disassembly
1	0000	0000	NOP
2	0002	0001	
3	0004	0002	
4	0006	0003	SLEEP
5	0008	0004	CLRWD
6	000A	0005	PUSH
7	000C	0006	POP
8	000E	0007	DAW
9	0010	0008	TBLRD
10	0012	0009	TBLRD
11	0014	000A	TBLRD
12	0016	000B	TBLRD
13	0018	000C	TBLWT
14	001A	000D	TBLWT
15	001C	000E	TBLWT
16	001E	000F	TBL
17	0020	0010	RET
18	0022	0011	RETF
19	0024	0012	RETUR
20	0026	0013	RETUR
21	0028	0014	
22	002A	0015	
23	002C	0016	
24	002E	0017	
25	0030	0018	
26	0032	0019	
27	0034	001A	
28	0036	001B	
29	0038	001C	
30	003A	001D	
31	003C	001E	
32	003E	001F	
33	0040	0020	
34	0042	0021	
35	0044	0022	
36	0046	0023	
37	0048	0024	

Opcode Hex | Machine | Symbolic

Configuration Bits

Address	Value	Category	Setting
300001	82	Oscillator	HS
		Fail-Safe Clock Monitor Enable	Disabled
		Internal External Switch Over Mode	Enabled
300002	0F	Power Up Timer	Disabled
		Brown Out Detect	Enabled
		Brown Out Voltage	2.0V
300003	1E	Watchdog Timer	Disabled-Controlled by SWDTEN bit
		Watchdog Postscaler	1:32768
300005	80	Master Clear Enable	MCLR enabled, RA5 input disabled
300006	85	Stack Overflow Reset	Enabled
		Low Voltage Program	Enabled
300008	00	Code Protect 000200-000FFF	Enabled
		Code Protect 001000-001FFF	Enabled
300009	40	Data EEPROM Code Protect	Enabled
		Code Protect Boot	Disabled
30000A	03	Table Write Protect 00200-00FFF	Disabled
		Table Write Protect 01000-01FFF	Disabled
30000B	E0	Data EEPROM Write Protect	Disabled
		Table Write Protect Boot	Disabled
		Config. Write Protect	Disabled
30000C	03	Table Read Protect 00200-00FFF	Disabled
		Table Read Protect 01000-01FFF	Disabled
30000D	40	Table Read Protect Boot	Disabled

MPLAB ICD 2 PIC18F1320 W:0 nov z dc c bank 0

My PIC programmer workspace, showing the device settings before erasure

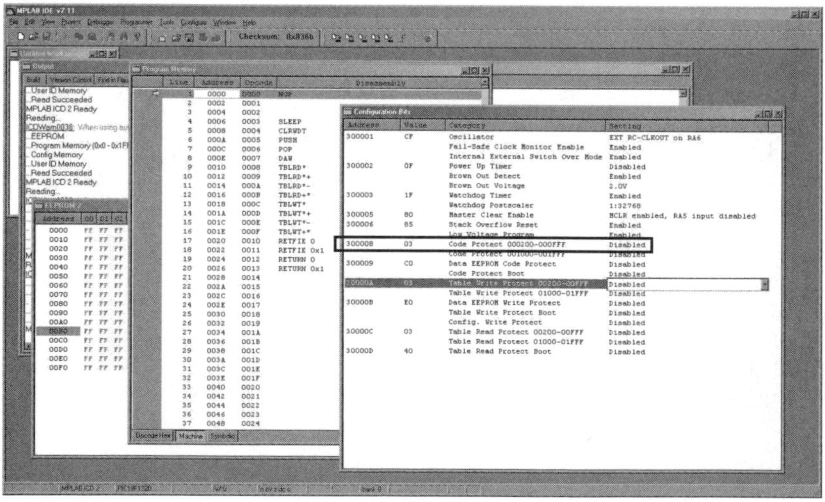

MPLAB IDE v7.11

File Edit View Project Debugger Programmer Tools Configure Window Help

Checksum: 0x836b

Output

Build | Version Control | Find in Files

...User ID Memory
...Read Succeeded
MPLAB ICD 2 Ready
Reading...
ICDWarn0038: When using aut
...EEPROM
...Program Memory (0x0 - 0x1FF
...Config Memory
...User ID Memory
...Read Succeeded
MPLAB ICD 2 Ready
Reading...

EEPROM

Address	00	01	02
0000	FF	FF	FF
0010	FF	FF	FF
0020	FF	FF	FF
0030	FF	FF	FF
0040	FF	FF	FF
0050	FF	FF	FF
0060	FF	FF	FF
0070	FF	FF	FF
0080	FF	FF	FF
0090	FF	FF	FF
00A0	FF	FF	FF
00B0	FF	FF	FF
00C0	FF	FF	FF
00D0	FF	FF	FF
00E0	FF	FF	FF
00F0	FF	FF	FF

Program Memory

Line	Address	Opcode	Disassembly
1	0000	0000	NOP
2	0002	0001	
3	0004	0002	
4	0006	0003	SLEEP
5	0008	0004	CLRWDT
6	000A	0005	PUSH
7	000C	0006	POP
8	000E	0007	DAW
9	0010	0008	TBLRD*
10	0012	0009	TBLRD*+
11	0014	000A	TBLRD*-
12	0016	000B	TBLRD+*
13	0018	000C	TBLWT*
14	001A	000D	TBLWT*+
15	001C	000E	TBLWT*-
16	001E	000F	TBLWT+*
17	0020	0010	RETFIE 0
18	0022	0011	RETFIE 0x1
19	0024	0012	RETURN 0
20	0026	0013	RETURN 0x1
21	0028	0014	
22	002A	0015	
23	002C	0016	
24	002E	0017	
25	0030	0018	
26	0032	0019	
27	0034	001A	
28	0036	001B	
29	0038	001C	
30	003A	001D	
31	003C	001E	
32	003E	001F	
33	0040	0020	
34	0042	0021	
35	0044	0022	
36	0046	0023	
37	0048	0024	

Opcode Hex | Machine | Symbolic

Configuration Bits

Address	Value	Category	Setting
300001	CF	Oscillator	EXT RC-CLKOUT on RA6
		Fail-Safe Clock Monitor Enable	Enabled
		Internal External Switch Over Mode	Enabled
300002	0F	Power Up Timer	Disabled
		Brown Out Detect	Enabled
		Brown Out Voltage	2.0V
300003	1F	Watchdog Timer	Enabled
		Watchdog Postscaler	1:32768
300005	80	Master Clear Enable	MCLR enabled, RA5 input disabled
300006	85	Stack Overflow Reset	Enabled
		Low Voltage Program	Enabled
300008	03	Code Protect 000200-000FFF	Disabled
		Code Protect 001000-001FFF	Disabled
300009	C0	Data EEPROM Code Protect	Disabled
		Code Protect Boot	Disabled
30000A	03	Table Write Protect 00200-00FFF	Disabled
		Table Write Protect 01000-01FFF	Disabled
30000B	E0	Data EEPROM Write Protect	Disabled
		Table Write Protect Boot	Disabled
		Config. Write Protect	Disabled
30000C	03	Table Read Protect 00200-00FFF	Disabled
		Table Read Protect 01000-01FFF	Disabled
30000D	40	Table Read Protect Boot	Disabled

MPLAB ICD 2 PIC18F1320 W:0 nov z dc c bank 0

The device settings after erasure

In the before shot, note the settings of the security fuses in the Configuration Bits window and the values programmed in the flash ROM, shown in the Program Memory window. In the after shot, the security fuses switch to being disabled, while the flash ROM contents in the Program Memory window read identically to what was programmed in previously. A different part of the code array was actually still erased, but I could probably have fixed that by cutting a bigger piece of electrical tape.

I've heard reports that since this hack was published, Microchip started putting metal shields over the code memory array as well as the fuses, making it a bit more difficult to pull off this trick. Still, this hack underscores the fact that quite often, the hardest part of silicon hacking is removing the outer package, and fortunately, there are cheap, if obscure, services available to assist with that problem.

HACKING SD CARDS

Years later, I found myself hacking into yet another interesting device with flash memory: an SD card. I'd already torn down SD cards when investigating a batch of potentially fake cards that found their way into Chumby production units, which I discuss in "Fake MicroSD Cards" on page 156. This time, my intent was to figure out how to get an SD card to do something it wasn't made to do. This particular hack was another team effort with my friend xobs, and it was funded by DARPA's Cyber Fast Track (CFT) initiative. The brainchild of uberhacker .mudge (one of the original crew of L0pht), CFT was a hack on the US government to make it smarter about innovation, particularly on matters related to internet security. We pulled it off around the same time we were working on Novena and I was collaborating with Jie Qi on Chibitronics.

xobs and I discovered that some SD cards contain vulnerabilities that allow arbitrary code execution on the memory cards themselves. We also found that similar classes of vulnerabilities exist in related devices like USB flash drives and solid-state drives. On the dark side, code execution on a memory card enables MITM attacks where the card seems to behave one way but in fact does something else as an attacker intercepts and manipulates communications between the card and the device using it. On the light side, however, this vulnerability also gives hardware enthusiasts access to a very cheap and ubiquitous source of microcontrollers.

Some of the eggs—or rather, SD cards—we cracked open to find the vulnerability

How SD Cards Work

To understand the hack, you need to know how SD cards are structured. The information I'm about to explain applies to all *managed flash* devices, which includes microSD, SD, and MMC, as well as the eMMC and iNAND devices typically soldered onto the mainboards of smartphones to store the operating system and other private user data.

Flash memory is billed as a contiguous, reliable storage medium, and it's really cheap—so cheap that the premise is literally too good to be true. In reality, all flash memory is riddled with defects, without exception. It crafts the illusion of reliability through sophisticated error correction and bad-block management functions. This system is the result of a constant arms race between the engineers and mother nature: every time the fabrication process shrinks transistors, memory becomes cheaper but more unreliable. Likewise, with every generation of chips, engineers create more sophisticated and complicated algorithms to compensate for nature's propensity for entropy and randomness at the atomic scale.

These algorithms are too complicated and too device-specific to be run at the application or operating system level, so every flash memory disk ships with a reasonably powerful microcontroller to run a custom set of disk abstraction algorithms. Even tiny microSD cards contain not one, but *at least* two, chips: a controller and at least one flash chip. (High-density cards stack multiple flash dies.)

Inside a microSD card. The small square in the upper-right corner is a microcontroller SoC mounted on top of the larger flash memory chip that it manages.

In my experience, the quality of the flash chip(s) integrated into memory cards varies widely. The chip could be anything from high-grade, factory-new silicon to material with more than 80 percent bad sectors. If you're concerned about e-waste, you may (or may not) be pleased to know that memory card vendors commonly use recycled flash chips salvaged from discarded parts. Larger vendors tend to offer more consistent quality, but even the largest players staunchly reserve the right to mix and match flash memory chips with different controllers yet sell the assembly as the same part number. That's a nightmare if you're dealing with implementation-specific bugs.

A memory card's embedded microcontroller is often a heavily modified Intel 8051 or ARM CPU that approaches 100 MHz performance levels and has several hardware accelerators on-die. Amazingly, adding these controllers to a memory card only costs about $0.15 to $0.30, particularly for companies that can fab both the flash memory and the controllers in the same business unit. Even more interestingly, due to the high cost of testing chips at the wafer level, it's probably net cheaper to add a microcontroller that manages bad blocks, rather than thoroughly test and characterize each raw flash memory chip. And in fact, managed flash devices tend to be cheaper per bit than raw flash chips, despite the extra functionality.

Every flash implementation has unique algorithmic requirements, multiplying the number of hardware abstraction layers a microcontroller must handle. This complexity inevitably leads to bugs, meaning indelibly burning a static body of code into on-chip ROM just isn't feasible, particularly for third-party controllers.

Thus, a firmware loading and update mechanism is virtually mandatory. End users are rarely exposed to this process since it all happens in the factory, but the mechanism exists. While exploring the electronics markets in China, I've seen shopkeepers burn firmware onto a card that "expands" the card's capacity. In other words, they load firmware that reports the capacity of a card as much larger than the actual available storage. The fact that this is possible at the point of sale indicates the update mechanism is likely not well secured.

Reverse Engineering the Card's Microcontroller

xobs and I discovered an example of this vulnerability while exploring memory cards using AppoTech's AX211 and AX215 microcontrollers. We discovered a simple "knock" sequence transmitted over manufacturer-reserved commands (a command named `CMD63` followed by the bytes `A`, `P`, `P`, `O`) that dropped the controller into a firmware loading mode. After receiving the knock sequence, the card accepted the next 512 bytes and ran the data as code.

NOTE *The AppoTech chips I describe here technically integrate sufficient functionality that in an academic sense, they're not mere microcontrollers; they're full SoCs. But it's just* weird *to me to refer to the AppoTech as an SoC, so I won't. It will always be a microcontroller to me!*

The AppoTech system on this particular memory card also used an 8051 microcontroller. From the knock sequence beachhead, we used a combination of analyzing code with IDA, the interactive disassembler, and *fuzzing* (that is, giving the microcontroller invalid or random input to see how it responds) to reverse engineer most of the 8051's function-specific registers. That allowed us to develop novel applications for the controller without the manufacturer's proprietary documentation. We did most of this work with the Novena laptop hardware I described in Chapter 7.

As I alluded at the beginning of this chapter, we developed several bespoke tools to help us reverse engineer the SD card. One of the more interesting tools we (and by we, I mean primarily xobs) made is an interactive REPL (read-evaluate-print-loop) shell for executing arbitrary code on the SD card. The following listing shows what that environment looks like.

```
root@bunnie-novena:~/ax211-code# ./ax211 -d debug.bin
FPGA hardware v1.26
Debug mode APPO response [6]: {0x3f 0x00 0xc1 0x04 0x17 0xab}
Result of factory mode: 0
00000000  0f 41 1f 0f 0f 0f ff ff                          |.A......|
Expected 0x00 0x00, got 0x0f 0x41
Loaded debugger
Locating fixup hooks... Done
AX211> help
List of available commands:
   hello  Make sure the card is there
    peek  Read an area of memory
    poke  Write to an area of memory
    jump  Jump to an area of memory
 dumprom  Dump all of ROM to a file
  memset  Set a range of memory to a single value
    null  Do nothing and return all zeroes
  disasm  Disassemble an area of memory
     ram  Manipulate internal RAM
     sfr  Manipulate special function registers
    nand  Operate on the NAND in some fashion
   extop  Execute an extended opcode on the chip
   reset  Reset the AX211 card
    help  Print this help
For more information on a specific command, type 'help [command]'
AX211> help disasm
Help for disasm:
Disassemble a number of bytes at the given offset.
Usage: disasm [address] [bytes]
AX211> disasm 0x200 16
.org 0x0200
        nop
        nop
        reti

        nop
        mov R7, A
        reti

        mov R7, A
        nop
        mov R7, A
        nop
        mov R7, A
        nop
```

From inside this environment, we could run programs in a debugger, get a list of available commands and what they did by entering `help`, and disassemble sections of code by entering `disasm`. Although it took a lot of time to develop an interactive tool with such a rich feature set, the effort quickly paid off because we could test complex hypotheses using automated fuzzing frameworks.

The code upload size was limited to 512 bytes, which meant we had to partition the REPL environment between the host Novena computer and the target device.* For example, disassembling a particular region of memory breaks down to a script executed on the host side that drives issue requests to the AX211 to dump the requested portion of memory, followed by the disassembly algorithm running on the host ARM CPU.

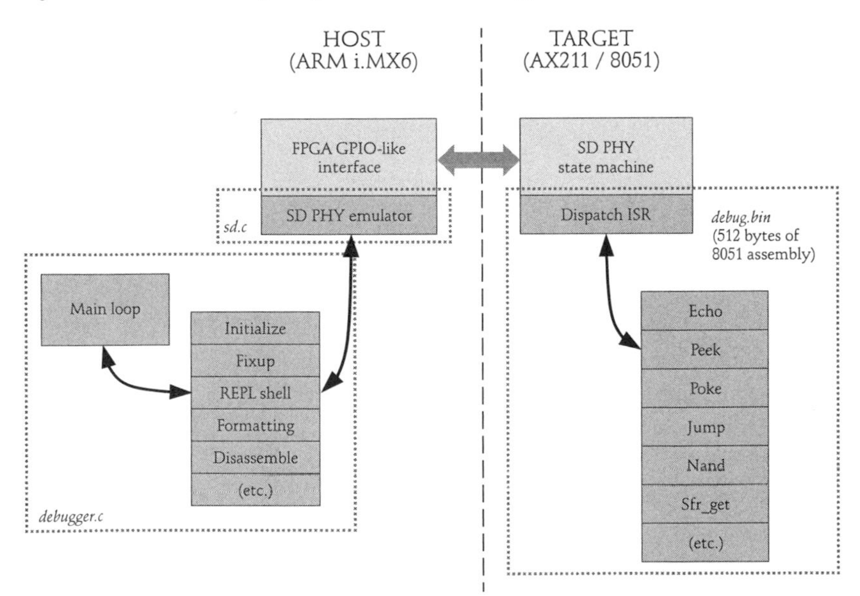

Partitioning the SD debugger functions between the host and the target

The tool we built started with an SD physical emulation layer, which I'll refer to as *PHY*. We used the FPGA built into the Novena to present a GPIO-like register API for the SD

* You can find a copy of the code at *https://github.com/xobs/ax211-code/*.

host PHY. There was one register for data output, one register for data input, and one register to bitwise set the data direction. The AX211 card was attached to the FPGA via a custom flex-circuit adapter.*

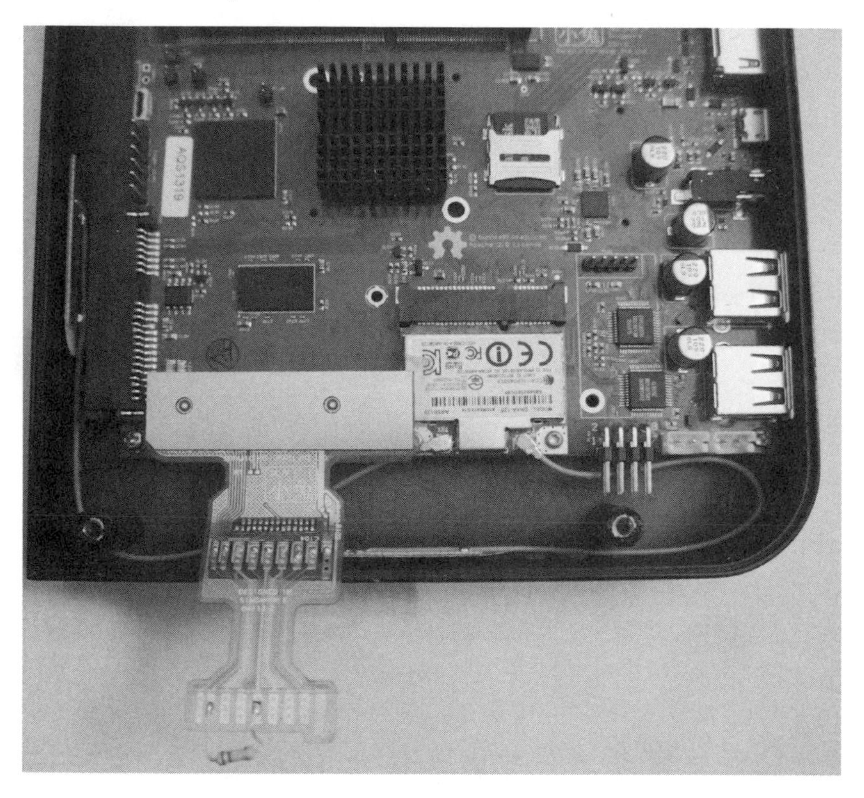

A flex-circuit adapter plugged into a Novena

The SD commands were received on the AX211 and processed by a hardware state machine attached to the embedded 8051 CPU. The state machine handled receiving the data, plus it computed and checked the cyclic redundancy code for error detection. Once a complete packet was received by the state machine, an interrupt notified the 8051 of the packet's arrival.

* Tangentially, we used the same flex adapter I mentioned in Chapter 8, which led in part to the development of Chibitronics.

We hijacked the interrupt processing mechanism and remapped the default handler to our own 512-byte code stub. That allowed us to define a novel set of SD commands that we used to implement the callback functions our REPL environment needed, like `peek`, `poke`, `jump`, NAND register manipulation, and so on. These callbacks were also an ideal hook for implementing an MITM attack.

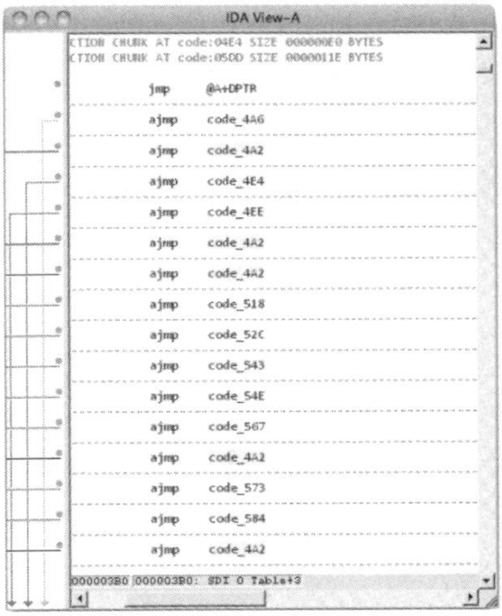

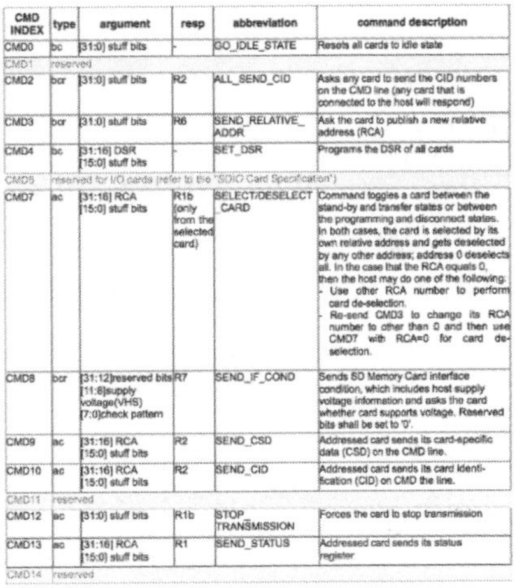

CMD INDEX	type	argument	resp	abbreviation	command description
CMD0	bc	[31:0] stuff bits	-	GO_IDLE_STATE	Resets all cards to idle state
CMD1	reserved				
CMD2	bcr	[31:0] stuff bits	R2	ALL_SEND_CID	Asks any card to send the CID numbers on the CMD line (any card that is connected to the host will respond)
CMD3	bcr	[31:0] stuff bits	R6	SEND_RELATIVE_ADDR	Ask the card to publish a new relative address (RCA)
CMD4	bc	[31:16] DSR [15:0] stuff bits	-	SET_DSR	Programs the DSR of all cards
CMD5	reserved for I/O cards (refer to the "SDIO Card Specification")				
CMD7	ac	[31:16] RCA [15:0] stuff bits	R1b (only from the selected card)	SELECT/DESELECT_CARD	Command toggles a card between the stand-by and transfer states or between the programming and disconnect states. In both cases, the card is selected by its own relative address and gets deselected by any other address; address 0 deselects all. In the case that the RCA equals 0, then the host may do one of the following: - Use other RCA number to perform card de-selection. - Re-send CMD3 to change its RCA number to other than 0 and then use CMD7 with RCA=0 for card de-selection.
CMD8	bcr	[31:12]reserved bits [11:8]supply voltage(VHS) [7:0]check pattern	R7	SEND_IF_COND	Sends SD Memory Card interface condition, which includes host supply voltage information and asks the card whether card supports voltage. Reserved bits shall be set to '0'.
CMD9	ac	[31:16] RCA [15:0] stuff bits	R2	SEND_CSD	Addressed card sends its card-specific data (CSD) on the CMD line.
CMD10	ac	[31:16] RCA [15:0] stuff bits	R2	SEND_CID	Addressed card sends its card identification (CID) on CMD the line.
CMD11	reserved				
CMD12	ac	[31:0] stuff bits	R1b	STOP_TRANSMISSION	Forces the card to stop transmission
CMD13	ac	[31:16] RCA [15:0] stuff bits	R1	SEND_STATUS	Addressed card sends its status register
CMD14	reserved				

The callback functions for the REPL, displayed in IDA

I don't know how many other manufacturers leave their firmware updating sequences unsecured. AppoTech is a relatively minor player in the SD controller world; a handful of companies that you've probably never heard of also produce SD controllers, including Alcor Micro, Skymedi, Phison, and SMI. Of course, there are also SanDisk and Samsung. Each has different mechanisms and methods for loading and updating firmware. But I know of at least one Samsung eMMC implementation using an ARM instruction set that had a bug requiring a firmware updater to be pushed to Android devices, indicating yet another potentially promising venue for further discovery.

Potential Security Issues

From a security perspective, our findings indicated that while memory cards look inert, they run code that could be modified to perform MITM attacks that are difficult to detect. There's no standard protocol or method to inspect and attest to the contents of the code running on the memory card's microcontroller. If you're using an SD card in a high-risk, high-sensitivity situation, don't assume that running a `security-erase` command (or some other secure erase tool) on a card will guarantee the complete erasure of sensitive data. If you really need data to disappear, I recommend disposing of your memory card through total physical destruction. Grind it up with a mortar and pestle if you have to.

A Resource for Hobbyists

From a DIY and hacker perspective, our findings suggested a potentially interesting source of cheap and powerful microcontrollers for use in simple projects. An Arduino clone—with an 8-bit, 16 MHz microcontroller—will set you back around $20. A microSD card with several gigabytes of memory and a microcontroller with several times the performance costs a fraction of the price. While SD cards are admittedly I/O-limited, some clever hacking of the microcontroller in an SD card could make for a very economical and compact data logging solution for I2C or SPI-based sensors.

HACKING HDCP-SECURED LINKS TO ALLOW CUSTOM OVERLAYS

"That's neat, but is it legal?" is a frequently asked question I get when hacking. Just as engineered systems have hacks, legal systems have loopholes. Some legal loopholes exist by design; others are unintentional. Either way, they can provide vital breathing room for innovation. When contemplating a hack, I consider legal issues as I do engineering constraints, similar to having to fit something within a case of a certain height or run for a certain length of time on a given battery.

Around 2011, when I was still at Chumby, we were puzzling about how to drive adoption in the face of the iPhone and Android phones consuming the market niche we hoped to occupy. Cost was an eternal barrier for user adoption, and the integral LCD in a chumby was by far the highest-cost item. Our then-CEO, Steve Tomlin, observed that the biggest screen in the house had yet to become connected to the internet in any meaningful way. And so this question was posed to me: could we find a way to kill two birds with one stone, removing the screen from our bill of materials while bringing TVs into the internet age? This was before products like the Google Chromecast or the Logitech Revue were introduced on the market.

It occurred to us that we could pack a cheap computer into a stick that plugs into an HDMI port. This solves the problem of getting chumby onto a TV screen, but then you're not watching your favorite movies or TV shows when the chumby is selected. We figured what people really wanted was some way to watch TV and have, say, Twitter or Facebook notifications pop up onscreen, too.

The concept is simple enough. Take the existing output from a cable box, Blu-ray player, or AV receiver; feed it into a box that blends in chumby content; and pass the resulting video on to a TV. But due to the ubiquitous application of HDCP encryption over digital video feeds, it is legally perilous to remix content if you do it the wrong way. Figuring out the right way to do it is how NeTV was born.

A NeTV sporting the Chumby logo

Inside the NeTV

Background and Context

NeTV was my response to the challenge of remixing existing video with internet content while staying within legal boundaries, aided by the public release of the master key to HDCP in September 2010. To help you understand this hack, let's start with a little background on HDCP.

High-bandwidth Digital Content Protection is a pixel-level encryption system used to encrypt video transmissions over HDMI. HDCP puts broadcasters and studios in control of the screens their content plays on, as those companies use the encryption as a copyright control mechanism. HDCP restricts legitimate content manipulation like picture-in-picture displays, content overlays, and third-party filtering and image modification. Combine HDCP with the DMCA, which criminalizes the circumvention of copyright control, and you'll realize that when watching certain videos, it's illegal to modify content on your own screen. That's why there are few HDMI video mixing solutions that actually operate on broadcast or movie content.

To recap, I had four goals for NeTV: enable consumer-side content remixing, allow users to eliminate ads or replace them with ads relevant to themselves, create an interactive TV experience, and make something compatible with any TV. To accomplish those goals, I designed NeTV as a man in the middle to take data from, say, a Blu-ray player, and apply the master key to give users a custom overlay. There are many applications for video overlays, but the basic scenario is that while you're enjoying content X, you'd also like to be aware of content Y. Combining the two content sources requires a video overlay mechanism.

With my MITM attack, NeTV overlaid a WebKit browser (the engine Safari and Chrome use) over any video feed. A concrete use case for this technology is overlaying Twitter feeds as news crawlers across a TV show to watch community commentary in real time on the same screen you're watching the show on. Some TV programs attempt to incorporate Twitter feeds already, but they've only done so on the source side; users can only watch hashtags the show displays. With this hack, however, the same broadcast program (say, a political debate) could have a very different viewing experience based on which hashtag is keyed into the viewer's Twitter crawler.

The simple fact that a trivial video overlay is an interesting topic illustrates the distortion of traditional rights and freedoms brought about by the DMCA. Unlike the HDCP strippers people speculated would come out of the master key's release, however, my hack never decrypted the original video data it operated on. Thus, it didn't circumvent copyright, and the DMCA couldn't apply to it. Loophole found!

How NeTV Worked

Of course, I released the exploit as an entirely open source project,* including the hardware and the Verilog implementation of the Spartan-6 FPGA I used to create the TMDS-compatible source and sink. TMDS is the signaling standard used by HDMI and DVI. The basic pipeline within the FPGA deserializes incoming video and reserializes it to the output. In this trivial mode, NeTV is simply a signal amplifier for the video: encrypted pixels in, encrypted pixels out—no decryption and no video manipulation.

NeTV could mix a user-generated content stream over an encrypted video feed because HDCP encrypts without validation. In other words, if a man in the middle tampers with the encrypted feed, the receiver simply accepts the tampered pixels as valid data, decrypts them, and presents them to the user. The lack of link verification is intentional and necessary. The natural bit error rate of HD video links is atrocious, but the human eye won't detect bit errors even on the level of 1 in every 10,000 bits. (At high error rates, users see a "sparkle" or "snow" on the screen, but the image is largely intact.) Allowing some pixel-level corruption keeps consumer costs low. Otherwise,

* You can read the documentation on the Sutajio Ko-Usagi wiki, although by the time of publication, the original NeTV product sold on Adafruit will probably have been phased out in favor of a newer, better implementation.

much higher-quality cables would be required along with FEC techniques to achieve a bit error rate compatible with strict cryptographic verification techniques like full-frame hashing.

Thus, NeTV's prime challenge is to derive a keystream identical and synchronized to the transmitter's keystream, encrypt the user-generated content with this keystream, and selectively swap the transmitter's pixels on the fly for user-encrypted pixels. If everything lines up, the receiver will decrypt an image that appears to be a perfect overlay of user-generated content on top of the original video feed.

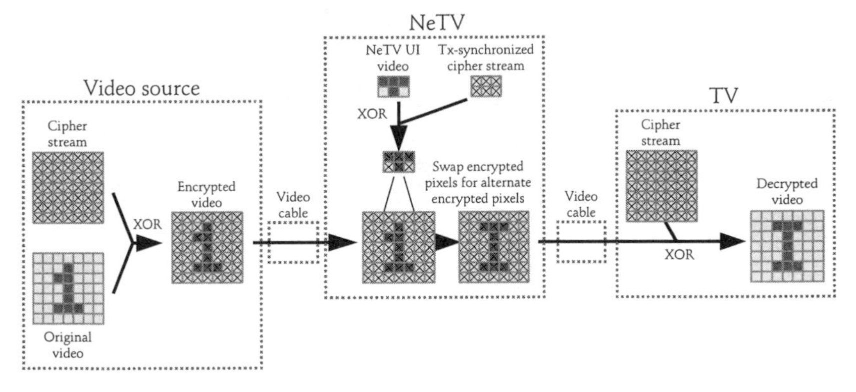

A high-level conceptual diagram of how NeTV worked

CREATING THE OVERLAY

To generate the user overlay content, we connected a tiny embedded Linux computer to an FPGA. From the Linux computer's standpoint, the FPGA emulates a parallel RGB LCD that you can access by using the frame buffer at */dev/fb0* (the filepath for the first frame buffer in Linux). The Linux computer would automatically launch a WebKit browser full-screen at boot, thus filling */dev/fb0* with the user's content.

The system selected which pixel to swap by observing the color of the WebKit overlay's video, a trick known as *chroma keying*. The overlay video wasn't encrypted and was generated by the user, so looking at the color of the overlay video was perfectly legal. Other more expressive and aesthetically appealing pixel-combining methods like alpha blending, however, would have required decrypting the original video, which would have been illegal.

If the overlay video matched a certain chroma key color (in this case, a specific shade of bright pink), the incoming video was displayed; otherwise, the overlay video was displayed. Following this system, users could create transparent "holes" in the custom UI to show the original video underneath. Since the UI was rendered by a WebKit browser, users could implement chroma keying by simply setting the background color in the CSS of the UI pages to that magic shade of pink. With those settings, the default state of a web page would be transparent, and all items rendered on top of it were opaque, so long as the UI elements avoided the chroma key color and turned off enhancements like anti-aliasing.

CRAFTING A KEYSTREAM

Of course, the chroma keying happened in the encrypted domain. Thus, the FPGA's second job was to snoop the HDMI link and craft a keystream identical to the transmitter's. First, the FPGA observed an I2C link found on HDMI known as the *data display channel* (DDC). The DDC enables monitors to report their capability records (called *extended display identification data,* or EDID) and is also where the encryption keys are exchanged.

By observing the key exchange handshake between the transmitter and the receiver, NeTV could mathematically extract the transmitter's and receiver's private keys with the help of the HDCP master key. Once the private-key vectors were derived, they could be multiplied exactly as they'd be in the source or sink to derive the shared secret, called Km. When that shared secret was written into the FPGA's HDCP engine, the cipher state was ready to go, allowing NeTV to encrypt overlays on the video transmitted between the video source and the video display device.

By considering legal constraints as just another engineering constraint, I was able to create a completely new device that proves a point: it's incorrect to automatically equate hacks that work around a DRM system with attempts to circumvent copyright. NeTV never decrypts previously encrypted video and can't operate without an existing, valid HDCP link, making it a bona fide, non-infringing, commercially useful application of the HDCP master key.

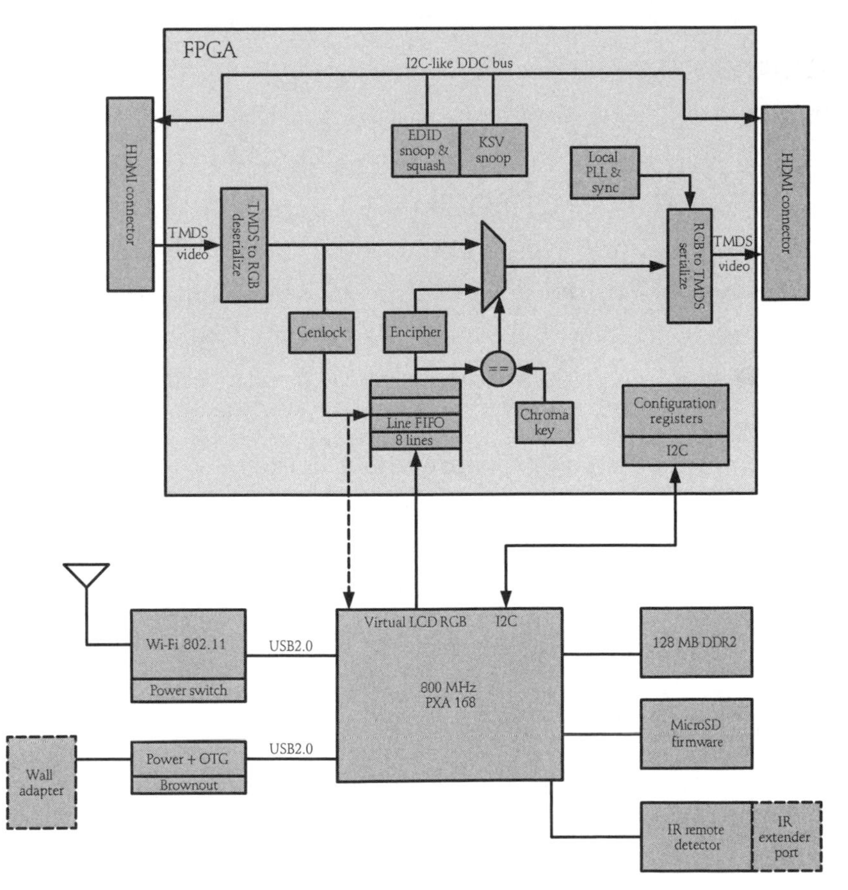

A more detailed block diagram showing how NeTV's FPGA worked

So far in this chapter, we've seen examples of different hardware hacking approaches and techniques, from physical penetration to system-level tool building and analysis to treating legal constraints as engineering problems. In "Who Are the Shanzhai?" on page 122, I discussed the legal approach of a project, codenamed Fernvale, to reverse engineer a mobile phone chipset. In addition to thinking about law as engineers, xobs and I had to pull out all the stops and apply every technical skill at our disposal to reverse engineer such a complex system. The rest of this chapter dives into some of these techniques.

HACKING A SHANZHAI PHONE

When xobs and I worked on Fernvale, our goal was to make a new platform derived from the hardware in my $12 gongkai phone and repatriate technical information into the open source IP system. We had no documentation whatsoever for some parts of the chip we wanted to reverse, but that didn't deter us. We navigated complex legal waters and created our own custom scripting language to program the chip's firmware to avoid subconscious plagiarism.

Compared to the firmware, though, the hardware reverse-engineering task was fairly straightforward. The documents we scavenged gave us a notion of the chip's pinout, and the pin naming scheme was sufficiently descriptive that I could apply common sense and experience to guess how to connect the chip. For ambiguous areas, I buzzed out some stripped-down phones with a multimeter or stared at them under a microscope to determine connectivity. In the worst cases, I'd probe a live phone with an oscilloscope to make sure I understood the connections correctly. The more difficult question was how to architect the hardware.

The System Architecture

We weren't gunning to build a phone, but rather something closer to Particle's Spark Core (since reborn as the Photon), a generic System-on-Module type of single-board computer built for Internet of Things applications. In fact, our original renderings and pinouts were designed to be compatible with the Spark ecosystem of hardware extensions, until we realized the gongkai phone's MT6260 microcontroller just had too many interesting peripherals to fit into such a small footprint.

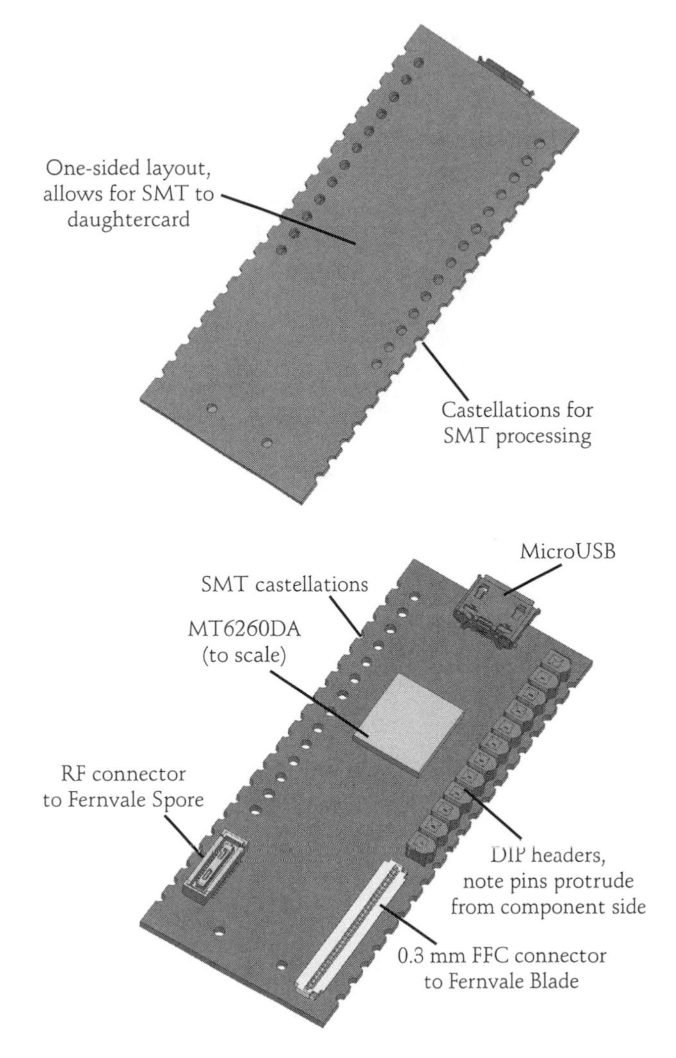

Early sketches of the Fernvale PCB

We settled eventually on a single-sided core PCB that we called the Fernvale Frond, which embedded the microUSB, microSD, battery, camera, speaker, and Bluetooth functionality (as well as the obligatory buttons and LED) on one board. The Frond turned out slim and small, at 3.5 mm thick, 57 mm long, and 35 mm wide. We included holes to mount a partial set of pin headers, spaced for Arduino compatibility, although the board could only be plugged into 3.3 V–compatible Arduino devices.

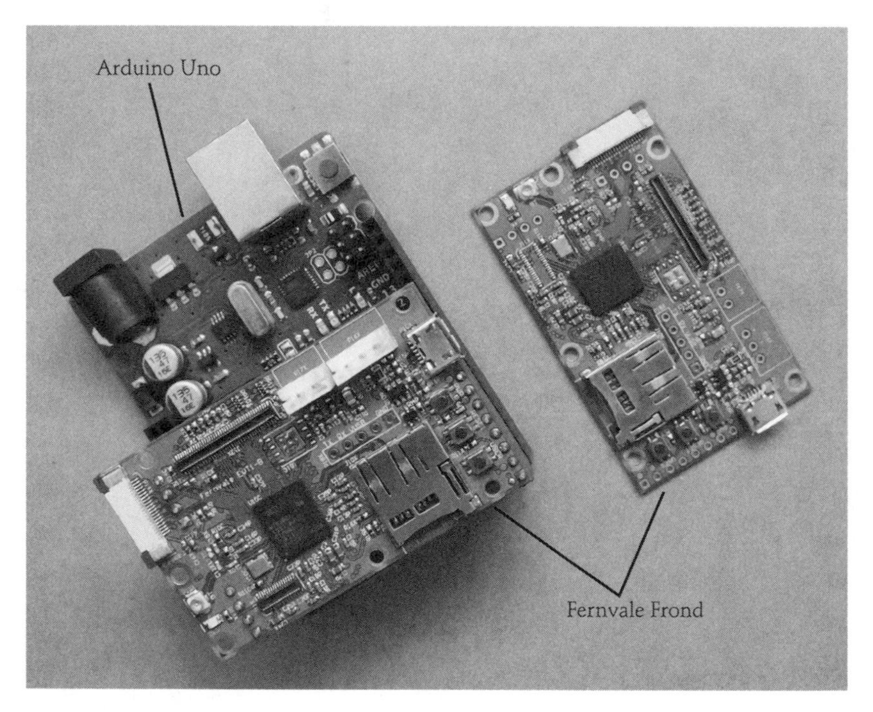

The actual implementation of the Fernvale Frond, pictured with an Arduino Uno for size reference

We broke the remaining peripherals out to a pair of connectors: one dedicated to GSM-related signals (GSM is the protocol for 2G cell phone networks) and the other to UI-related peripherals. We called the GSM board the Fernvale Spore and the UI board the Fernvale Blade. We split GSM into a module with many choices for the RF frontend to make GSM a bona fide user-installed feature, thus pushing the regulatory and emissions issue down to the user level. Splitting the UI-related features out to another board also reduced the cost of the core module and let users try the Frond in numerous scenarios without being locked into a particular LCD or button arrangement.

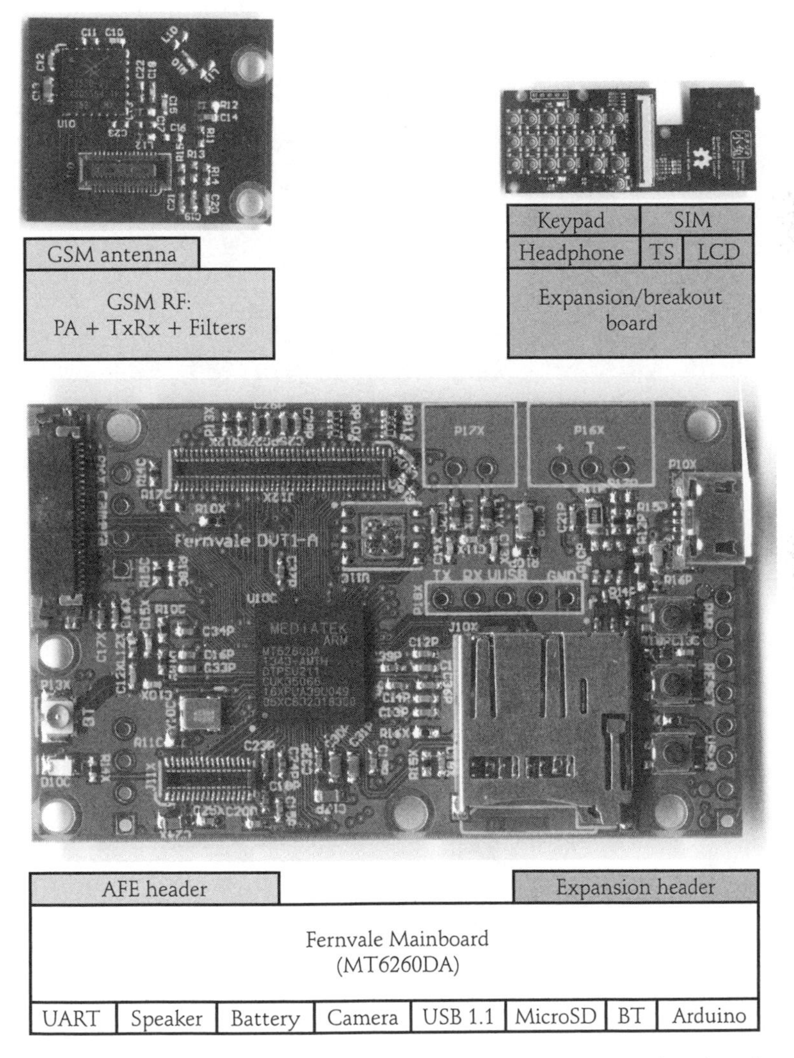

A Fernvale system diagram, showing the features of each of the three boards

Inside the MT6260

I had some X-rays taken of the MT6260 to help us identify fake components. We had to source our MT6260s on the gray market, and we wanted to guard against being sold empty epoxy blocks or remarked versions of other chips. The MT6260 has -DA and -A variants, where the difference is how much on-chip flash memory is included.

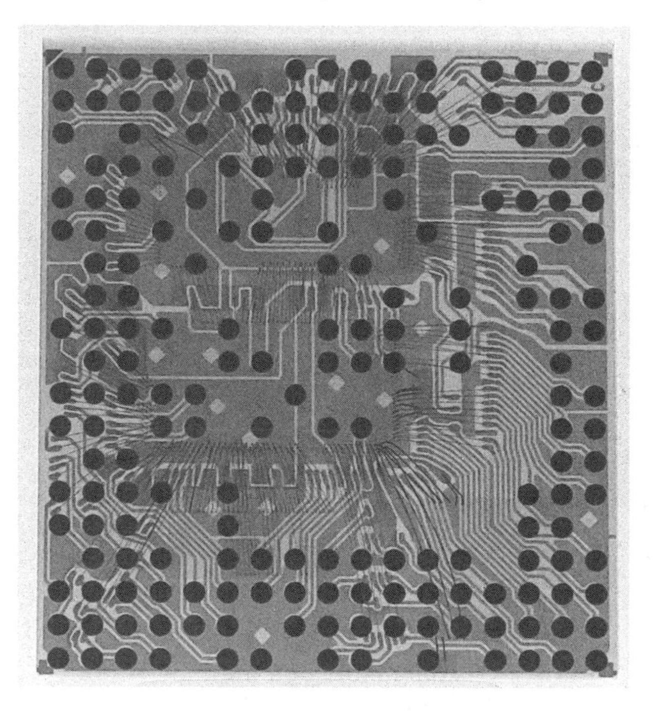

An X-ray of the MT6260 chip.
Look carefully to spot outlines of multiple ICs among the wire bonds.

To our surprise, this $3 chip didn't contain a single IC, but rather a set of at least four (possibly five) chips integrated into a single multichip module (MCM) containing hundreds of wire bonds. I remember back when the Pentium Pro's dual-die package came out in the late 1990s. It sparked arguments over yield costs of MCMs versus using a single big die; generally, MCMs were considered exotic and expensive.

I also remember at the same time Krste Asanović, then a professor at the MIT Artificial Intelligence Lab and later at UC Berkeley, told me that the future of electronics wasn't system-on-a-chip devices, but rather "system-mostly-on-a-chip" devices. The root of his claim was that the economics of adding in mask layers to merge DRAM, flash, analog, RF, and digital into a single process wasn't favorable; bonding multiple dies together into a single package was cheaper and easier.

It's still a race between the cost impact (in terms of both the per-unit cost and nonrecurring engineering costs) of adding more process steps in the semiconductor fab, and the yield impact, relative reworkability, and lower nonrecurring engineering cost of assembling modules. Single-chip, System-on-Chip devices were the zeitgeist when Krste made that observation and they still kind of are, so it was interesting to see a significant data point validating his insight.

Understanding the internal structure of the chip was also helpful in reverse engineering the system. Knowing that MediaTek was simply combining several chips together in a single package shed much-needed light on the purpose and organization of their APIs. It also tipped us off that certain elements of the system would be reused across several product categories and generations, so we knew we could draw meaningful conclusions from documentation on older or related chips. When you're piecing together a puzzle this complex, every clue helps, including those gained by just looking at the physical structure of the chip.

Reverse Engineering the Boot Structure

Shanzhai engineers in China seem to have access to just enough documentation to assemble a phone and customize its UI, but not enough to do a full OS port. After looking at enough phones, I eventually realized that all phones based on a particular chipset will have the same backdoor codes, and their GUIs are often inconsistent with the implemented hardware. For example, the $12 phone I tore down in Chapter 4 prompted me to plug headphones into the headphone jack for the FM radio to work, yet it has no headphone jack.

To make Fernvale accessible to engineers in the West through open source licensing, we had to reconstruct everything from scratch, including the toolchain, the firmware flashing tool, the OS, and the applications. But all the Chinese phone implementations simply relied on MediaTek's proprietary toolchain, meaning we had to do some reverse engineering to figure out the boot process and firmware upload protocol.

My first step in reversing a chip is always to dump the ROM, if possible. We found exactly one phone model with an external ROM that we could desolder (it used the -D ROMless variant of the chip), and we read its data using a conventional ROM reader. We saw very little ciphertext in the ROM, but there was a lot of compressed data. Here is a page from our notes after we did a static analysis on the ROM image:

```
0x0000_0000         media signature "SF_BOOT"
0x0000_0200         bootloader signature "BRLYT", "BBBB"
0x0000_0800         sector header 1 ("MMM.8")
0x0000_09BC         reset vector table
0x0000_0A10         start of ARM32 instructions
                      - stage 1 bootloader?
0x0000_3400         sector header 2 ("MMM.8")
                      - stage 2 bootloader?
0x0000_A518         thunk table of some type
0x0000_B704         end of code (padding until next sector)
0x0001_0000         sector header 3( "MMM.8") - kernel?
0x0001_0368         jump table + runtime setup (stack, etc.)
0x0001_0828         ARM thumb code start - possibly also
                      baseband code
0x0007_2F04         code end
0x0007_2F05         begin padding "DFFF"
0x0009_F005        end padding "DFFF"
0x0009_F006         code section begin "Accelerated
                      Technology / ATI / Nucleus PLUS"
0x000A_2C1A         code section end; pad with zeros
0x000A_328C         region of compressed/unknown data begin
0x007E_E200         modified FAT partition #1
0x007E_F400         modified FAT partition #2
```

The hexadecimal numbers on the left are memory addresses, and the text on the right describes what xobs and I thought was stored at each address. One concern about reverse engineering an SoC is it has an internal boot ROM that always runs before code is loaded from an external device. That internal ROM can also have signature and security checks that prevent tampering with the external code.

To determine how hard reverse engineering this system would be, we wanted to quickly figure out how much code was running inside the CPU before jumping to external boot code. A Tek MDO4104B-6 oscilloscope let us accomplish that task in just a couple of hours.

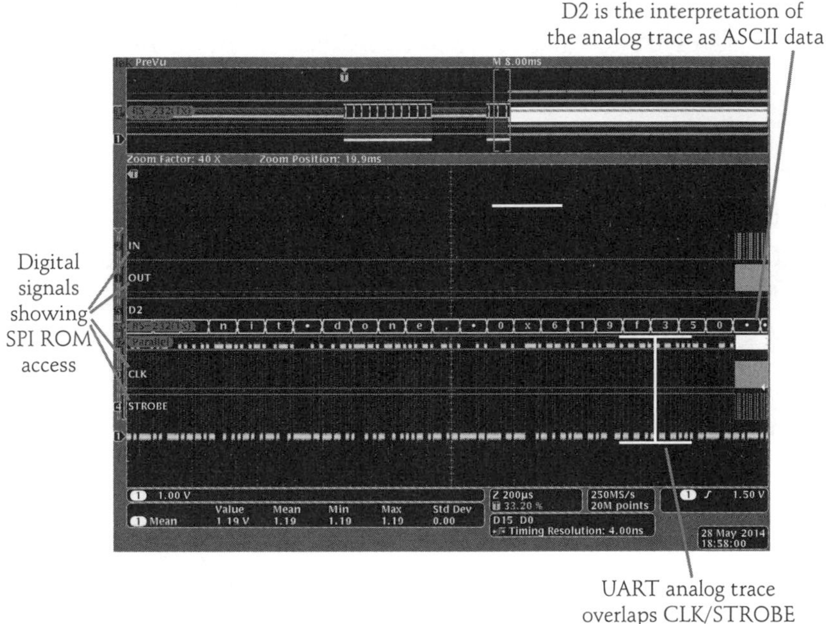
D2 is the interpretation of the analog trace as ASCII data
Digital signals showing SPI ROM access
UART analog trace overlaps CLK/STROBE from SPI ROM

Screenshot from the Tek MDO4104B-6.
The top quarter shows a zoomed-out view of the entire capture.
Notice how the SPI ROM accesses are punctuated with console output.

This particular oscilloscope has the uncanny ability to perform post-capture analysis on deep, high-resolution analog traces and output the result as digital data. For example, we could simply probe around the chip with a multimeter while cycling power until we saw something that looked like an RS-232 encoded signal, and then run a post-capture analysis to extract any ASCII text that was coded in the analog traces. Likewise, if we captured SPI traces, the oscilloscope could extract ROM access patterns through a similar method. By looking at the timing of text emissions versus SPI ROM address patterns, we quickly determined that if the internal boot ROM did any verification, it was minimal and nothing approaching the computational complexity of RSA encryption.

From there, we needed to speed up our measure-modify-test loop. Desoldering the ROM, sticking it in a burner, and resoldering it to the board were going to get old really fast. Fortunately, we'd implemented a NAND flash ROM emulator (we lovingly shortened that to ROMulator) on Novena, which we previously used to reverse engineer the AX211 contained in certain SD cards. We just reused that codebase and made an SPI ROMulator. We hacked up a GPBB and its corresponding FPGA code to add the ability to swap between the original boot SPI ROM and a dual-ported 64kiB emulator region that was also memory-mapped into the Novena Linux host's address space. Then, we plugged the phone into the laptop and put the ROMulator to work.

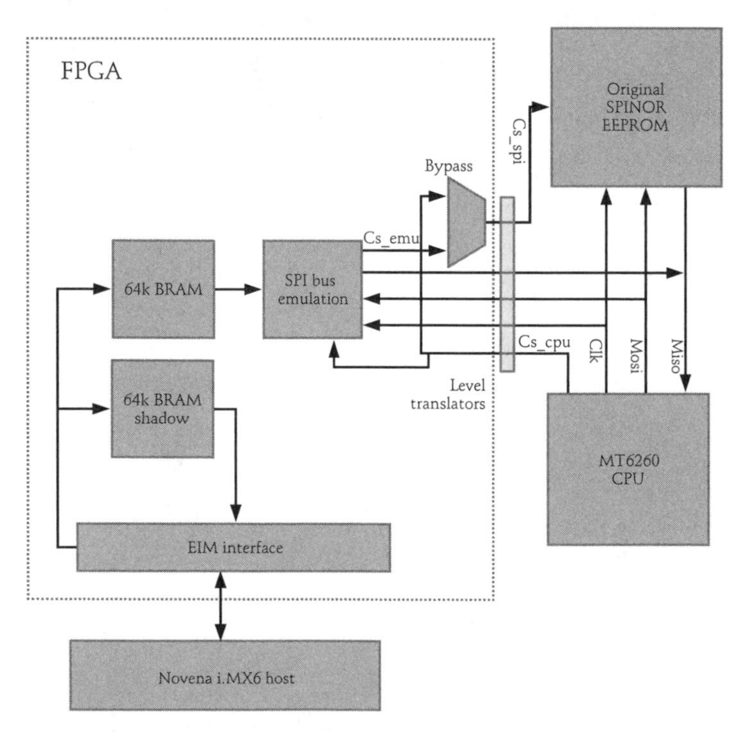

A block diagram of the SPI ROMulator FPGA

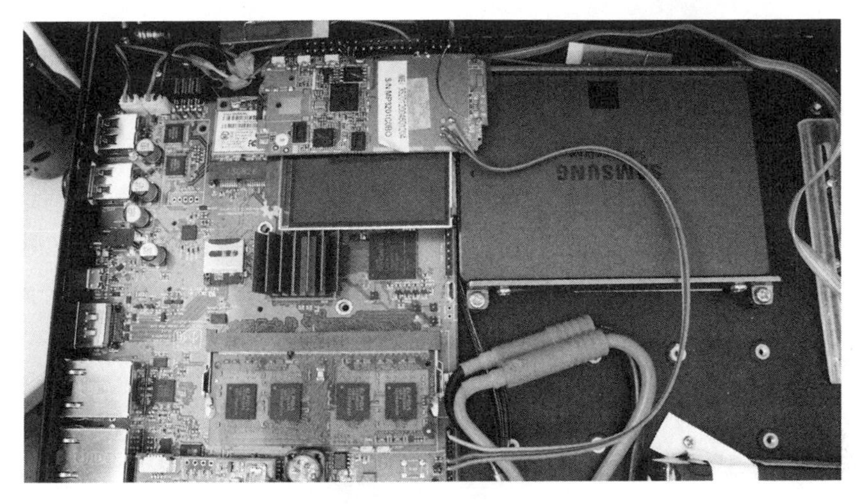

There's a phone in my Novena! What's that doing there?

With the address stream determined by the Tek oscilloscope, some rapid ROM patching by the ROMulator, and hints of a SHA-1 function existing in the ROM via a static code analysis using IDA, we determined that the initial bootloader (which we called the 1bl), was hash-checked using a SHA-1 appendix.

NOTE *The assembly for a hash function tends to have a very distinctive shape, or set of instructions, and a given hash also has some amount of magic numbers unique to it. Given those facts, when trying to reverse an authentication method, one of the first things a hacker does is use IDA to search for such constants near a function with the shape of the hash function in question.*

Building a Beachhead

The next step was to create a small interactive shell we could use as a beachhead for running experiments on the target hardware. Just as he did for the SD card reverse engineering project, xobs created a compact REPL environment, called Fernly, that supported commands like peeking at memory, writing data, and dumping CPU registers.

Designing the ROMulator to make the emulated ROM appear as a 64kiB memory-mapped window on a Linux host enabled useful POSIX abstractions like the `mmap()` function, the `open()` function (via */dev/mem*), the `read()` function, and the `write()` function to access the emulated ROM. xobs used these abstractions to create an I/O target for radare2, a portable reverse engineering framework. The I/O target automatically updated the SHA-1 hash every time we made changes in the 1bl code space. With that system in place, we could do cute things like interactively patch and disassemble code within the emulated ROM space.

```
bunnie@bunnie-novena-laptop: ~/code/radare2                    bunnie@bunnie-noven

0x00000c6b  0x0019046b 0x22f9b1f0 0x09200100 0xd0f000a1  k......"..  .....
0x00000c7b  0xfd0020fe 0x70eae4f7 0x0008f4bd 0x00100300  . .....p........
0x00000c8b  0x0086f800 0x00668070 0x0086b470 0x0085d070  ....p.f.p...p...
0x00000c9b  0x6f6f6670 0x6f742064 0x616d6f79 0x202c616d  pfood toyomama,
           0x00000c6b    6b04          lsls r3, r5, 17
           0x00000c6d    1900          movs r1, r3
           0x00000c6f    f0b1          cbz r0, 0x00000caf
           0x00000c71    f922          movs r2, 249
           0x00000c73    0001          lsls r0, r0, 4
           0x00000c75    2009          lsrs r0, r4, 4
           0x00000c77    a100          lsls r1, r4, 2
       ,=< 0x00000c79    f0d0          beq.n 0x00000c5d ;[1]
       |   0x00000c7b    fe20          movs r0, 254
       |   0x00000c7d    00fdf7e4      stc2 4, cr14, [r0, -988] ; 0xfffffc24
       |   0x00000c81    ea70          strb r2, [r5, 3]
       |   0x00000c83    bdf40800      ; <UNDEFINED> 0xf4bd0008 ;[2]
       |     0xffffffff84bdc97() ; hit1_0
       |   0x00000c87    0003          lsls r0, r0, 12
       |   0x00000c89    1000          movs r0, r2
       |   0x00000c8b    00f88600      strb.w r0, [r0, r6]
       |   0x00000c8f    7080          strh r0, [r6, 2]
       |   0x00000c91    6600          lsls r6, r4, 1
       |   0x00000c93    70b4          push {r4, r5, r6}
       |   0x00000c95    8600          lsls r6, r0, 2
      ,==< 0x00000c97    70d0          beq.n 0x00000d7b ;[3]
      ||   0x00000c99    8500          lsls r5, r0, 2
      ||   0x00000c9b    7066          str r0, [r6, 100]
      ||   0x00000c9d    6f6f          ldr r7, [r5, 116]
      ||   0x00000c9f    6420          movs r0, 100
      ||   0x00000ca1    746f          ldr r4, [r6, 116]
      ||   ;-- hit1_0:
      ||   0x00000ca3    796f          ldr r1, [r7, 116]
      ||   0x00000ca5    6d61          str r5, [r5, 20]
      ||   0x00000ca7    6d61          str r5, [r5, 20]
      ||   0x00000ca9    2c20          movs r0, 44
      ||   0x00000cab    2578          ldrb r5, [r4, 0]
      ||   0x00000cad    0a0d          lsrs r2, r1, 20
      ||   0x00000caf    00f8b500      strb.w r0, [r0, r5, lsl 3]
      ||   0x00000cb3    26f64335      ; <UNDEFINED> 0xf6263543 ;[4]
      ||     0xffffffff862773d() ; hit1_0
      ||   0x00000cb7    0000          movs r0, r0
      ||   0x00000cb9    f0a2          add r2, pc, 960 ; (adr r2, 0x0000107c)
      ||   0x00000cbb    f925          movs r5, 249
      ||   0x00000cbd    4841          adcs r0, r1
```

Patching some code in the ROM

We also wired up the power switch of the phone to an FPGA I/O. That allowed us to write automated scripts that toggled the power on the phone while updating the ROM contents so we could automatically fuzz unknown hardware blocks.

Attaching a Debugger

We had to take an unconventional approach to attach a debugger to the code in the ROM, because locating critical blocks was difficult, and JTAG was multiplexed with critical functions on the target device. xobs emulated the ARM core and used his Fernly shell to reflect virtual loads and stores to the live target. We were able to attach a remote debugger to the emulated core that way, bypassing the need for JTAG entirely. That also let us use cross-platform tools like IDA on x86 for the reversing UI.

At the heart of this debugging technique was QEMU, a multiplatform system emulator. QEMU supports emulating ARM targets, specifically the ARMv5 chip our target device used. We made a new virtual machine type, called Fernvale, that implemented part of the observed hardware on the target and simply passed unknown memory accesses directly to the device.

The Fernly shell was stripped down to support only three commands: write, read, and zero-memory. The write command pokes a byte, word, or dword of data into RAM on the live target. A read command reads a byte, word, or dword from the live target. The zero-memory command is an optimization, as the operating system writes large quantities of zeros across a large memory area.

We also hooked and emulated the serial port registers, allowing a host system to display serial data as if it were printed on the target device. Finally, we emulated SPI, IRAM, and PSRAM as they'd appear on the real device. Other areas of memory were either trapped and funneled to the actual device or left unmapped and reported as errors by QEMU.

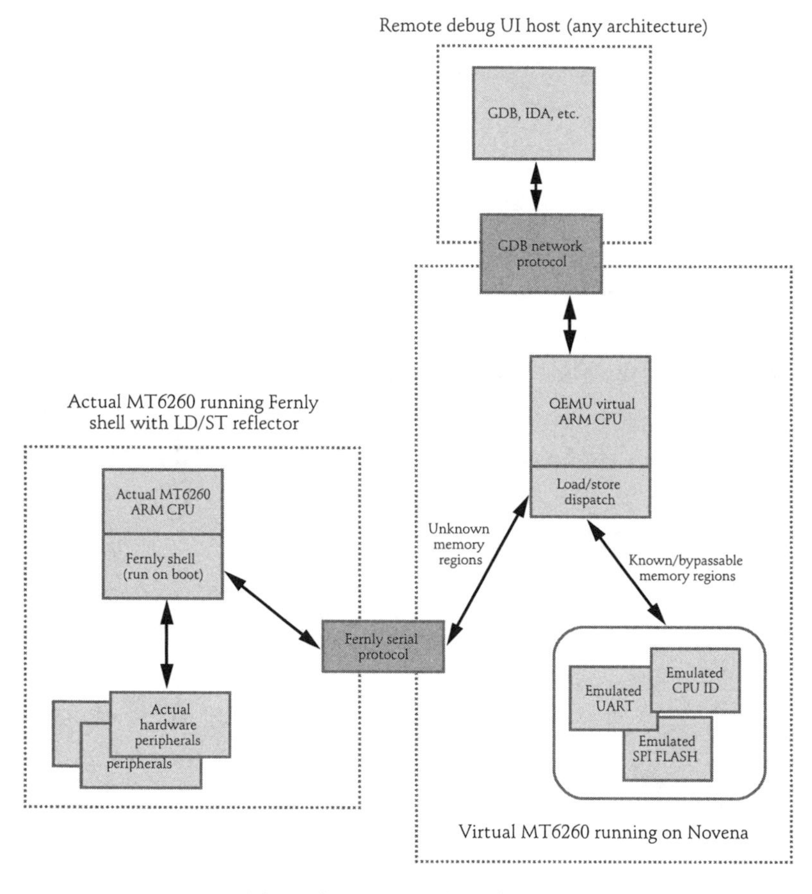

The architecture of the debugger

Invoking the debugger was a multistage process. First, we primed the actual MT6260 target with the Fernly shell environment. Then, we booted the QEMU virtual ARM CPU with a version of the original vendor image primed with a known register state at a convenient point in the boot process. At this point, code execution proceeded on the virtual machine until a load or store was performed to an unknown address. On that load or store, virtual machine execution paused while a query was sent to the real MT6260 via the Fernly shell interface. The load or store was then executed on the real machine, which would relay the results of the load or store to the virtual machine so execution could resume.

We couldn't run Fernly directly from the SPI ROM because the vendor binary's initialization routine modified SPI ROM timings. But of course Fernly would have crashed if a store happened to land somewhere inside its memory footprint. To avoid the possibility of a load or store overwriting the Fernly shell code, we hid the code in a region of IRAM that was trapped and emulated. Emulating the target CPU let us attach a remote debugger like IDA via GDB over TCP. The debugger had complete control over the emulated CPU and could access its emulated RAM. Here is an example of the output of the hybrid QEMU/live-target debug harness.

```
bunnie@bunnie-novena-laptop:~/code/fernvale-qemu$ ./run.sh

~~~ Welcome to MTK Bootloader V005 (since 2005) ~~~
**==================================================**

READ WORD Fernvale Live 0xa0010328 = 0x0000... ok
WRITE WORD Fernvale Live 0xa0010328 = 0x0800... ok
READ WORD Fernvale Live 0xa0010230 = 0x0001... ok
WRITE WORD Fernvale Live 0xa0010230 = 0x0001... ok
READ DWORD Fernvale Live 0xa0020c80 = 0x11111011... ok
WRITE DWORD Fernvale Live 0xa0020c80 = 0x11111011... ok
READ DWORD Fernvale Live 0xa0020c90 = 0x11111111... ok
WRITE DWORD Fernvale Live 0xa0020c90 = 0x11111111... ok
READ WORD Fernvale Live 0xa0020b10 = 0x3f34... ok
WRITE WORD Fernvale Live 0xa0020b10 = 0x3f34... ok
```

This output shows the trapped serial writes appearing on the console, plus a log of thc writes and reads executed by the emulated ARM CPU as they were relayed to the live target running the reduced Fernly shell. This was our beachhead.

From there, xobs and I discovered the offsets of a few IP blocks that were reused from previous known MediaTek chips by searching for their "signature" in memory. A signature could be as simple as the power-on default register values, or something more complex, like changes in bit patterns due to the side effects of bit set or clear registers located at offsets within the IP block's address space. Following the signatures helped us find the register offsets of several peripherals and generate a memory map.

Starting Address	Ending Address	Size of Region	Description
0x00000000	0x0fffffff	0x0fffffff	PSRAM map, repeated and mirrored at 0x00800000 offsets
0x10000000	0x1fffffff	0x0fffffff	Memory-mapped SPI chip
??????????	??????????	??????????	????????????????????????????????
0x70000000	0x7000cfff	0xcfff	On-chip SRAM (maybe cache?)
??????????	??????????	??????????	????????????????????????????????
0x80000000	0x80000008	0x08	Config block (chip version, etc.)
0x82200000	??????????	??????????	
0x83000000	??????????	??????????	
0xa0000000	0xa0000008	0x08	Config block (mirror?)
0x10010000	??????????	??????????	(?SPI mode?) ????????????????????
0x10020000	0xa0020e10	0x0e10	GPIO control block
0xa0030000	0xa0030040	0x40	WDT block + 0x08 -> WDT register (?) + 0x18 -> Boot src (?)
0xa0030800	??????????	??????????	????????????????????????????????
0xa0040000	??????????	??????????	????????????????????????????????
0xa0050000	??????????	??????????	????????????????????????????????
0xa0060000	??????????	??????????	?? Possible IRQs at 0xa0060200 ??
0xa0070000	==========	==========	== Empty (all zeroes) ===========
0xa0080000	0xa008005c	0x5c	UART1 block
0xa0090000	0xa009005c	0x5c	UART2 block
0xa00a0000	??????????	??????????	????????????????????????????????

This memory map shows what content is stored at different address ranges on the chip. For instance, the second address range in the map (0x10000000 to 0x1FFFFFFF) consisted of 0x0FFFFFFF bytes corresponding to a memory-mapped SPI chip.

Booting an OS

After finding the register offsets, we progressed rapidly on many fronts, but our goal (to port NuttX, a BSD-based real-time operating system, to the device) remained elusive. There was no documentation on the interrupt controller within the canon of shanzhai datasheets. We found the routines that installed the interrupt handlers through static analysis of the binaries, but we couldn't determine the address offsets of the interrupt controller itself.

All we could do was open the MediaTek codebase and refer to the header file that contained the register offsets and bit definitions of the interrupt controller. This fit within our self-imposed limitations to not breach copyright, because facts are not copyrightable. I describe the legal reasoning behind this idea in Chapter 4, under "Dealing with Copyrights" on page 138. After looking up those facts, we created our own custom scripting language, called Scriptic, to avoid unconsciously plagiarizing anything from the existing codebase.

Building a New Toolchain

Requiring users to own a Novena ROMulator to hack on Fernvale wasn't a scalable solution, however. To round out the story, we created a complete developer toolchain. The compiler was fairly cut-and-dried; many standard compilers support ARM as a target, including clang and GCC. But making open tools for flashing the MT6260 was much trickier. All the existing tools we knew supported the protocol version required by the MT6260 were proprietary Windows programs. That meant we had to reverse engineer the MediaTek flashing protocol and write our own open source tool.

Fortunately, a blank, unfused MT6260 shows up as */dev/ttyUSB0* when you plug it into a Linux host. In other words, it shows up as an emulated serial device over USB. That took care of the lower-level details of sending and receiving bytes to the device, leaving us to reverse engineer the protocol layer.

xobs located the internal boot ROM of the MT6260 and performed static code analysis to learn more about the protocol. He also did some static analysis on MediaTek's flashing tool and captured live traces using a USB protocol analyzer to clarify the remaining details. Here is a summary of the commands he extracted, as we used in our open version of the USB flashing tool.

```
enum mtk_commands {
  mtk_cmd_old_write16 = 0xa1,
  mtk_cmd_old_read16 = 0xa2,
  mtk_checksum16 = 0xa4,
  mtk_remap_before_jump_to_da = 0xa7,
  mtk_jump_to_da = 0xa8,
  mtk_send_da = 0xad,
  mtk_jump_to_maui = 0xb7,
  mtk_get_version = 0xb8,
  mtk_close_usb_and_reset = 0xb9,
  mtk_cmd_new_read16 = 0xd0,
  mtk_cmd_new_read32 = 0xd1,
  mtk_cmd_new_write16 = 0xd2,
  mtk_cmd_new_write32 = 0xd4,
  // mtk_jump_to_da = 0xd5,
  mtk_jump_to_bl = 0xd6,
  mtk_get_sec_conf = 0xd8,
  mtk_send_cert = 0xe0,
  mtk_get_me = 0xe1, /* Responds with 22 bytes */
  mtk_send_auth = 0xe2,
  mtk_sla_flow = 0xe3,
  mtk_send_root_cert = 0xe5,
  mtk_do_security = 0xfe,
  mtk_firmware_version = 0xff,
};
```

This is just a C `enum` structure, making it a very geeky way of specifying a mapping of numbers to command meanings. For example, `mtk_cmd_old_write16` is command 0xA1, `mtk_command_old_read16` is command 0xA2, and so on.

Fernvale Results

After about a year of on-and-off effort between work on the Novena and Chibitronics campaigns, we were able to boot a port of NuttX on the MT6260, supporting a minimal set of hardware peripherals. It was enough for us to roughly reproduce the functionality of an AVR used in an Arduino-like context, but not much more.

xobs and I presented our results at the 31st Chaos Communication Congress (CCC), and events actually took an unexpected twist as we wrote our proposal. The week before submission, we learned that MediaTek released the LinkIT ONE development platform, based on the MT2502A, in conjunction with Seeed Studios. The LinkIT ONE is an Internet of Things platform made for entrepreneurs and hobbyists. It's integrated into the Arduino framework and features an open API that enables the full functionality of the chip, including GSM functions. But the core OS that boots on the MT2502A in the LinkIT ONE is still proprietary, and you can't access the hardware without going through the API calls provided by the Arduino shim.

Realistically, it's still going to be a while before we can port a reasonable fraction of the MT6260's features into the open source domain. It's quite possible we'll never be able to do a blob-free implementation of the GSM call functions, as those are controlled by a DSP unit that's even more obscure and undocumented than the MT6260. Given the robust functionality of the LinkIT ONE compared to Fernvale, we decided to leave the question of whether there was value in continuing the effort to reverse engineer the MT6260 to the open source community. In the end, there was a lot of enthusiasm for the project, but not a lot of action. The LinkIT ONE's introduction took a lot of wind out of the sails of the Fernvale project, which has since been effectively retired.

This is, in fact, the fate of most open source projects. There are dozens, if not hundreds, of open source operating systems but only one Linux. The truth is that there are far more interesting ideas than capable developers to execute them. For an open source project to catch fire and become self-sustaining, it has to not only pass the minimum viable product (MVP) stage but also meet a receptive audience with a real need for the project. Sometimes your project strikes a chord, and a huge community pushes it forward. Other times, you get a lot of nice, helpful onlookers who nod appreciatively but are unwilling or too busy with day jobs to jump in. And still other times, you yell into a void or, worse, get torn to shreds on some internet forum about how flawed and pointless your project is.

CLOSING THOUGHTS

Given the nature of open source projects, I tend to take a page from my startup days and follow a "fail forward fast" philosophy. Try a bunch of different things, see what sticks, learn from your mistakes, and try again. It's important not to get too wedded to any one idea, especially if the idea isn't working out. Finally, you'll find it helps to be more about the journey than the destination. Fernvale was most certainly an epic journey; xobs and I learned a lot, honed a set of tools and skills that we continue to use to this day for other projects, and most importantly, had a lot of fun.

In the next chapter, we'll take a look at another kind of hacking that will become increasingly relevant to all of us over the coming decades—that of biological systems.

10. biology and bioinformatics

I once came across a beautiful diagram in *Science** showing the metabolic pathways of one of the smallest bacteria, *Mycoplasma pneumoniae*. It reminded me of staring at an Apple II schematic when I was less than a decade old. Back then, I knew that the Apple II schematic's fascinatingly complex mass of lines was a map to the computer in front of me, though I didn't know quite enough to do anything with that map. But the point was that a map existed, so despite its imposing appearance, it gave me hope that I could unravel such complexities. Biological "schematics" like the one on the next page give me the same hope.

* Eva Yus et al., "Impact of Genome Reduction on Bacterial Metabolism and Its Regulation," *Science* 326, no. 5957 (2009): 1263–1268, *http://science.sciencemag.org/content/326/5957/1263/*.

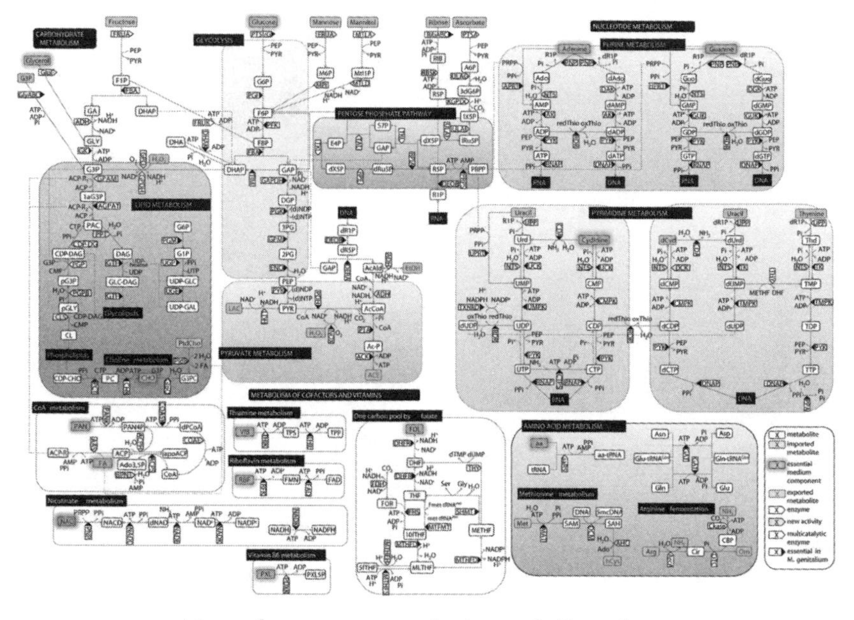

Mycoplasma pneumoniae*'s metabolic pathway*

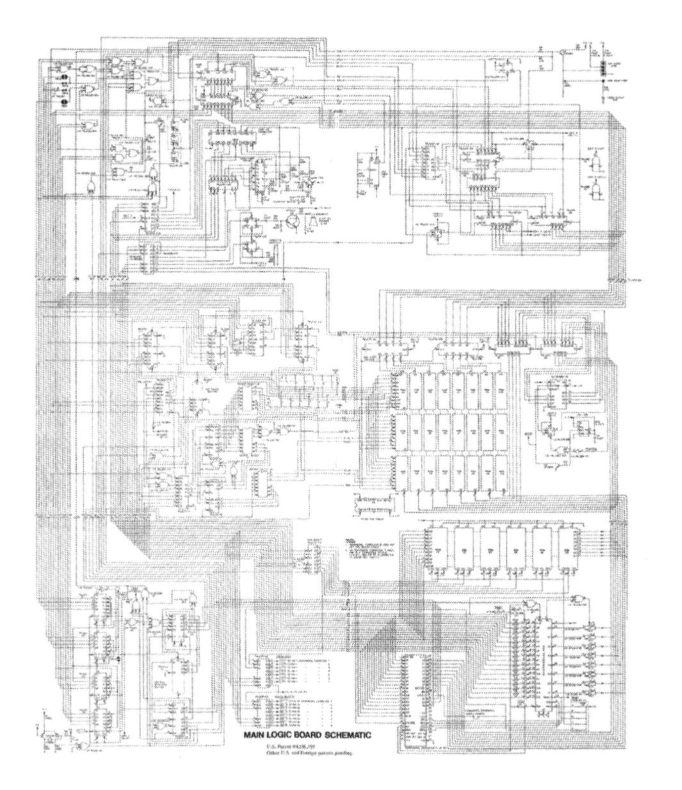

The Apple II schematic from my wall

The *M. pneumoniae* diagram isn't quite as precise as the Apple II schematic, but from 10,000 feet, they feel similar in complexity and detail. The metabolic diagram is detailed enough for me to trace a path from glucose to ethanol, and the Apple II schematic is detailed enough for me to trace a path from the CPU to the speaker. And just as a biologist wouldn't make much of a box with 74LS74 attached to it, an electrical engineer wouldn't make much of a box with ADH inside it. (A 74LS74 contains two instances of a synchronous electronic storage device, and ADH is alcohol dehydrogenase, an enzyme coded by gene *MPN564* that can turn acetaldehyde into ethanol.)

Furthering the computer analogy, though, the *Science* article's authors also included a list that read like a BOM for *M. pneumoniae* in their supplemental material. The pentagonal boxes in the diagram are *enzymes*, proteins that catalyze specific chemical reactions. Each enzyme is listed with a functional description along with its gene sequence, which is equivalent to source code.

At the very end of that list, I saw a table of uncharacterized genes. If you've done a bit of reverse engineering, you've probably made similar tables for parts or function calls in an electronic system. They're the first place I go for fresh clues when I get stuck. I find it heartening to see biologists and hackers applying similar techniques to reverse engineering complex systems.

COMPARING H1N1 TO A COMPUTER VIRUS

The comparison of biological systems to computer systems doesn't stop at the metabolic level. I once read a fascinating article in *Nature** that compared the pathogenic components

* Gabriele Neumann, Takeshi Noda, and Yoshihiro Kawaoka, "Emergence and Pandemic Potential of Swine-Origin H1N1 Influenza Virus," *Nature* 459, no. 7249 (2009): 931–939, *http://www.nature.com/nature/journal/v459/n7249/full/nature08157.html.*

of the *novel H1N1 virus* (better known as *swine flu*) to those of other flu strains, and that article got me thinking about how digital and organic viruses compare. For example, how big is an organic virus relative to a digital one? To put the question another way, how many bits does it take to kill a human, or at least make one quite sick? In exploring this idea, I found it helpful to draw a few analogies between the digital and organic worlds.

DNA and RNA as Bits

When the H1N1 pandemic broke out in 2009, the virus was comprehensively sequenced and logged in the National Center for Biotechnology Information's (NCBI) Influenza Virus Resource database, and the data collected there is amazing. I love the specificity of the records. For example, the entire sequence of an instance of influenza known as A/Italy/49/2009(H1N1) isolated from the nose of a 26-year-old female *Homo sapiens* returning from the United States to Italy is on the NCBI website. Here are the first 120 bits of the DNA sequence:

```
atgaaggcaa tactagtagt tctgctatat acatttgcaa ccgcaaatgc agacacatta
```

With 120 bits total, each symbol (A, T, G, or C) represents 2 bits of information. In genes, this can be alternatively represented as an amino acid sequence, where every three DNA symbols are a *codon* corresponding to one amino acid. Long chains of *amino acids* fold into complex structures called *proteins* that give structure and function to a cell, and chains of amino acids too short to be a complete protein are often called *peptides*. Using a translation lookup table that biologists call the standard genetic code, I converted the previous sequence into the following peptide: `MKAILVVLLYTFATANADTL`.

In this sequence, each symbol represents an amino acid, which is the equivalent of six bits or three DNA bases per amino acid. There are 20 amino acids in the canonical codon table, and each letter corresponds to a different amino acid. M is methionine, K is lysine, A is alanine, and so on.

Now, consider RNA, which passes information from DNA on how to synthesize proteins to the rest of the cell. As with DNA, each base in RNA specifies one of four possible symbols (in this case, A, U, G, or C), so a single base corresponds to two bits of information. DNA and RNA are information-equivalent on a one-to-one mapping. Think of DNA as a program stored on disk and RNA as the same program loaded into RAM. When DNA is loaded, protein synthesis instructions are transcribed into RNA, but all T bases are replaced with U bases.

Proteins, then, are the output of running an RNA program. Proteins are synthesized according to the instructions in RNA on a three-to-one mapping. You can think of proteins like pixels in a frame buffer, as follows:

- A complete protein is like an image on the screen.
- Each amino acid on a protein is like a pixel.
- Each pixel has a depth of six bits, due to the three-to-one mapping of a medium that stores two bits per base.
- Finally, each pixel goes through a color palette (the codon translation table) to transform the raw data into a final rendered color. Unlike a computer frame buffer, however, different biological proteins vary in amino acid count (analogous to a pixel count).

To ground this in a specific example, imagine that six bits stored as ATG on your hard drive (DNA) are loaded into RAM (RNA) as AUG because T is transcribed as U when going from DNA to RNA. When the RNA program in RAM is executed, AUG is translated to a pixel (amino acid) of color M, or methionine, which is the biological “start” codon—that is, the first instruction in every valid RNA program.

As a shorthand, since DNA and RNA are one-to-one equivalent, bioinformaticists represent gene sequences in DNA format, even if the biological mechanism is in RNA format. The influenza virus has an RNA architecture, rather than DNA, and the 120 bits of DNA I showed earlier correspond to an RNA subroutine in influenza. That subroutine codes for the HA gene, which produces an H1 variety of the hemagglutinin protein. This is the *H1* in the H1N1 designation of swine flu.

Organisms Have Unique Access Ports

Given that background information, if you think of organisms as computers with IP addresses, each functional group of cells in the organism listens to the environment through its own active port. As port 25 maps specifically to SMTP services on a computer, port H1 maps specifically to the windpipe region on a human. Interestingly, the same port H1 maps to the intestinal tract on a bird. Thus, the same H1N1 virus will attack the respiratory system of a human and the gut of a bird. In contrast, H5—the variety of hemagglutinin protein found in H5N1, the deadly avian flu—specifies the port for your inner lungs. As a result, H5N1 is much deadlier than H1N1 because it attacks your inner lung tissue, causing severe pneumonia. H1N1 is less deadly because it attacks a more benign port that just makes you blow your nose a lot and cough up loogies.

NOTE *Researchers are still discovering more about the H5 port. The* Nature *article I read indicated that perhaps certain human mutants have lungs that don't listen on the H5 port. People whose lungs ignore the H5 port would have a better chance of surviving an avian flu infection, while those that open port H5 on the lungs have no chance to survive (make your time . . . all your base pairs are belong to H5N1).**

* If you're not familiar with this turn of phrase, see *https://en.wikipedia.org/wiki/All_your_base_are_belong_to_us*.

Knowing a virus is deadly, you can figure out how many bits it takes to kill a human (or at least make one quite sick) by calculating the number of bits in the viral genome. The question, then, is how many bits are in this instance of H1N1? The raw number of bits, by my count, is 26,022; the number of actual coding bits is approximately 25,054. I say "approximately" because in some places, the virus does the equivalent of self-modifying code to create two proteins out of a single gene. It's hard to say what counts as code and what counts as an incidental, nonexecuting NOP sled required for the self-modified code.

That means it takes about 25Kb or 3.2KB of data to code for a virus that has a nontrivial chance of killing a human. This is more efficient than a computer virus like MyDoom, which comes in around 22KB. Knowing that I could be killed by 3.2KB of genetic data is humbling. Then again, with roughly 800MB of data in my genome, there's bound to be an exploit or two.

Hacking Swine Flu

One interesting consequence of reading this *Nature* article and having access to the virus sequence is that in theory, I now know how to modify the virus sequence to make it deadlier. For instance, the *Nature* article notes that variants of the PB2 influenza gene with glutamic acid at position 627 in the sequence have a *low pathogenicity*, meaning they aren't very deadly. However, PB2 variants with lysine at the same position increase the likelihood of mortality.

Let's see the sequence of PB2 for H1N1. Going back to the NCBI database, I found the following amino acid sequences around position 627:

```
601 QQMRDVLGTFDTVQIIKLLP
621 FAAAPPEQSRMQFSSLTVNV
641 RGSGLRILVRGNSPVFNYNK
```

The numbers to the left indicate the position of the first symbol in each line of the sequence; I'll follow that convention for the rest of this discussion. Check the line labeled 621, and note the E in position 627. E is the symbol for glutamic acid. Thankfully, H1N1 seems to be a less-deadly version of influenza; perhaps this is why fewer people died from contracting H1N1 than the media might have led you to believe.

Now, let's reverse this back to the DNA code:

```
621   F   A   A   A   P   P   E   Q   S   R
1861 ttt gct gct gct cca cca gaa cag agt agg
```

Notice the GAA codes for E. To modify this genome to be deadlier, you'd simply need to replace GAA with one of the codes for lysine (K). Lysine can have a code of either AAA or AAG. Thus, a deadlier variant of H1N1 would have a coding sequence like this:

```
621   F   A   A   A   P   P   K   Q   S   R
1861 ttt gct gct gct cca cca aaa cag agt agg
                             ^ changed
```

So, a single base-pair change—simply flipping two bits—might be all you'd need to turn the H1N1 swine flu virus into a deadlier variant. Theoretically, I could apply a series of well-known biological procedures to synthesize this strain and actually implement the hack. As a first step, I could go to a DNA synthesis website and order the modified sequence to get my deadly little project going for just over $1,000. Some of those companies have screening procedures to protect against DNA sequences that could be used to implement biohazardous products, but even if they happened to screen for HA variants, there are well-known protocols for site-directed mutagenesis that could possibly be used to modify a single base of RNA from material extracted from normal H1N1.

Adaptable Influenza

Of course, I have to give influenza some credit. It packs a deadly punch in 3.2KB, and despite scientists' best efforts, we haven't eradicated it. Could influenza do hacks like the one I just described on its own already?

The short answer is yes.

In fact, the influenza virus evolved to allow for these adaptations. Normally, when DNA is copied, an error-checking protein runs over the copied genome to verify that no mistakes were made. This keeps the error rate quite low. But remember, the influenza virus uses an RNA architecture. It therefore needs a different mechanism from DNA for copying.

Inside its protein capsule, the influenza virus packs code for a protein complex called *RNA-dependent RNA polymerase*, which is a tiny machine for copying RNA off of RNA templates. Normally, RNA is only generated by transcribing DNA, not by copying an existing piece of RNA, so this mechanism is essential for the replication of RNA-based influenza. Significantly, RNA-dependent RNA polymerase omits an error-checking protein that would prevent mutations. The result is that influenza makes about one error per 10,000 base pairs that get copied. The influenza genome is about 13,000 base pairs long, so on average, every copy of an influenza virus has one random mutation.

Some of these mutations make no difference; others render the virus harmless; and quite possibly, some render the virus much more dangerous. Since viruses are replicated and distributed in astronomical quantities, the chance that this little hack could end up occurring naturally is in fact quite high. I think this is part of the reason health officials were so worried about H1N1: people had no resistance to it, and even though it wasn't as deadly as it could have been, the strain was probably just a couple of mutations away from being a much bigger health problem.

There is one other important subtlety to the RNA architecture of the influenza virus, aside from its high mutation rate: the virus's genetic information is stored as eight separate, relatively short, snippets of RNA. In many other viruses and simple organisms, genetic information is instead stored as a single unbroken strand.

To understand why that's important, consider what happens when a host is infected by two types of the influenza virus at the same time. If the genes were stored as a single piece of DNA or RNA, there would be little opportunity for the genes between the two types to shuffle. But because influenza stores its genes as eight separate snippets, those genes mix freely inside the infected cell and are randomly shuffled into virus packets as they emerge. If you're unlucky enough to get two types of flu at once, the result is a potentially novel strain of flu, as RNA strands are copied, mixed, picked out of the metaphorical hat, and then packed into virus particles. This process is elegant in that the same mechanism allows for mixing of an arbitrary number of strains in a single host. If you can infect a cell with three or four types of influenza at once, the result is an even wilder variation of flu particles.

This mechanism is part of the reason novel H1N1 is called a *triple-reassortant* virus. Through a series of dual infections or perhaps a single calamitous infection of multiple flu varieties, novel H1N1 acquired a mix of RNA snippets that gave it high transmission rates and made it something humans weren't innately immune to. That's the perfect storm for a pandemic.

If there were a computer analogy to this RNA-shuffling model, it would be a virus that distributes itself in the form of unlinked object code files plus a small helper program that, upon infecting a host, relinks its files in a random order before copying and redistributing itself. It would also search for similar viruses that may already be infecting that computer and on occasion link in object code with matching function templates from the other viruses. This rearrangement and novel relinking of the code itself would foil classes of antivirus software that search for virus signatures based on fixed code patterns. It would also proliferate a diverse set of viruses in the wild, with less predictable properties.

The influenza virus's multilevel adaptation mechanism is remarkable. The virus has both a slowly evolving point mutation mechanism and a mechanism for drastically altering its properties in a single generation through gene-level mixing with other viruses. It doesn't work quite like sex, but the result is probably just as good, if not better. It's also remarkable that these two important properties of the virus arise as a consequence of using RNA instead of DNA as the genetic storage medium.

A Silver Lining

Since there are so many variants of flu, no vaccine can target all types of the virus, but the H1N1 story does have a silver lining. Apparently, a patient who contracted swine flu during the pandemic created a novel antibody with the remarkable ability to confer immunity to all 16 subtypes of influenza A. A group of researchers sifted through the patient's white blood cells and managed to isolate four B cells that contained the code to produce this antibody. They cloned the cells and produced antibodies, facilitating further research into a potential vaccine that could confer broad protection against the flu.

I found this really interesting at a gut level because it gives me hope that if a killer virus did wipe out most of humanity, maybe a small group of people would survive it.

REVERSE ENGINEERING SUPERBUGS

In 2011, a "superbug" strain of *E. coli* (a species of bacteria with subtypes that can cause food poisoning) called EHEC O104:H4 broke out in Europe. When I found out that scientists at BGI, located in Shenzhen, had released the entire sequence of O104:H4 freely online for anyone to examine, I got very curious about the situation. I couldn't help but wonder exactly what tools bioinformaticists use to analyze DNA sequences. Manually inspecting the relatively simple sequences of the influenza virus is one thing, but there must be computational tools to help make sense of more complicated organisms like *E. coli*.

Fortunately, my perlfriend (`s/perl/girl/`) is also a noted bioinformaticist. She took some time out of her busy schedule to show me some tools of the trade. It turns out most of the tools for analyzing DNA are freely available online. Since DNA is just sequences of A's, T's, G's, and C's, the standard data interchange format is plain old ASCII text, which means you can do a lot of analysis using command-line tools like `grep`, `sed`, and `awk`.

The O104:H4 DNA Sequence

The raw sequence data BGI provided was a set of oversampled subsequences that we needed to assemble by matching up overlapping regions. Stitching subsequences together is a bit like composing a large picture from small photos taken at random. With enough sampling, you'll eventually create a mostly complete picture, but the image will still have ambiguities, particularly in areas with regular patterns.

The genome of O104:H4 was provided as a list of over 500,000 short DNA samples. The assembly process stitched the short DNA samples together into 513 contiguous fragments of DNA (known as *contigs*), with a total genome length of 5.3 million base pairs. An organism like *E. coli* has just one big loop of DNA, so there were 513 spots where limitations in the sequencing technology (or just bad luck) missed an unknown number of base pairs, preventing us from knowing the entire, unbroken sequence. Notably, a typical, non-superbug strain of

E. coli has around 4.6 million base pairs, so O104:H4 is probably at least 15 percent longer. Likewise, this strain would take more time to replicate than a non-drug-resistant strain. Take a look at contig 34 of the assembly:

```
AAATGGTATTCCTGTTCACGATACTATTGCCAGAGTTGTATCCTGTATCAGTCCTGC
AAAATTTCATGAGTGCTTTATTAACTGGATGCGTGACTGCCATTCTTCAGATGATAA
AGACGTCATTGCAATTGATGGAAAAACGCTCCGGCACTCTTATGACAAGAGTCGCCG
CAGGGGAGCGATTCATGTCATTAGTGCGTTCTCAACAATGCACAGTCTGGTCATCGG
ACAGATCAAGACGGATGAGAAATCTAATGAGATTACAGCTATCCCAGAACTTCTTAA
CATGCTGGATATTAAAGGAAAAATCATCACAACTGATGCGATGGGTTGCCAGAAAGA
TATTGCAGAGAAGATACAAAAACAGGGAGGTGATTATTTATTCGCGGTAAAAGGAAA
CCAGGGGCGGCTAAATAAAGCCTTTGAGGAAAAATTTCCGCTGAAAGAATTAAATAA
TCCAGAGCATGACAGTTACGCAATTAGTGAAAAGAGTCACGGCAGAGAAGAAA
```

I could have picked any contig, and it probably would have made about as much sense to you as this block of letters. Aside from making gratuitous pop culture references (the word *GATTACA* occurs 252 times in the genome of O104:H4), the raw DNA sequence isn't very insightful. It's a bit like staring at binary machine code. To analyze the data, you need to "decompile" the "methods" contained within the code.

In this case, we were searching for DNA sequences that code for *proteins*. As I mentioned earlier, proteins are complex, often interwoven chains of molecules consisting of small building blocks known as amino acids. Cells get things done using proteins: some proteins turn sugar into energy, others use that energy to move around or change the cell's shape, and still others are responsible for copying and repairing the cell.

Fortunately, protein sequences are highly conserved in DNA. Nature tends to reuse protein structures, with few modifications, between organisms. Thus, a function that has been determined through a biological experiment, even on another species, can often be correlated with a sequence of DNA. For instance, one common experiment for determining the function of a sequence is to cut a piece of DNA out of a cell and observe what happens to the cell; the loss of function resulting from the missing DNA is often indicative of the protein's role in the cell.

Biologists have amassed decades of research on what certain proteins do into huge databases. Thus, to figure out what a chunk of DNA means, you can do a fuzzy pattern match between your DNA of interest and the database of known proteins.

Reversing Tools for Biology

I needed two tools to reverse engineer DNA: a protein database and a piece of software called BLASTX. Both are free to download online.

THE UNIPROT DATABASE

I downloaded a list of known proteins from the Universal Protein Resource, or UniProt (*http://www.uniprot.org/*). In 2011, a search of the database for "drug resistance" restricted to *E. coli* organisms yielded a list of 1,378 proteins that scientists have identified over the years as parts of the *E. coli* bacteria's drug-resistance machinery. Every year, new discoveries are added to the database.

Here's a snippet from the database that describes a protein that gives O104:H4 resistance to a drug you may recognize:

```
>sp|P0AD65|PBP2_ECOLI Penicillin-binding protein 2
OS=Escherichia coli (strain K12) GN=mrdA PE=3 SV=1

MKLQNSFRDYTAESALFVRRALVAFLGILLLTGVLIANLYNLQIVRFTDYQTRSNENRIK
LVPIAPSRGIIYDRNGIPLALNRTIYQIEMMPEKVDNVQQTLDALRSVVDLTDDDIAAFR
KERARSHRFTSIPVKTNLTEVQVARFAVNQYRFPGVEVKGYKRRYYPYGSALTHVIGYVS
KINDKDVERLNNDGKLANYAATHDIGKLGIERYYEDVLHGQTGYEEVEVNNRGRVIRQLK
EVPPQAGHDIYLTLDLKLQQYIETLLAGSRAAVVVTDPRTGGVLALVSTPSYDPNLFVDG
ISSKDYSALLNDPNTPLVNRATQGVYPPASTVKPYVAVSALSAGVITRNTTLFDPGWWQL
PGSEKRYRDWKKWGHGRLNVTRSLEESADTFFYQVAYDMGIDRLSEWMGKFGYGHYTGID
LAEERSGNMPTREWKQKRFKKPWYQGDTIPVGIGQGYWTATPIQMSKALMILINDGIVKV
PHLLMSTAEDGKQVPWVQPHEPPVGDIHSGYWELAKDGMYGVANRPNGTAHKYFASAPYK
IAAKSGTAQVFGLKANETYNAHKIAERLRDHKLMTAFAPYNNPQVAVAMILENGGAGPAV
GTLMRQILDHIMLGDNNTDLPAENPAVAAAEDH
```

PBP2_ECOLI* is linked to penicillin resistance and is a mutated gene that determines the shape of the bacteria. It seems this resistant variant adapted to operate despite the presence of penicillin; bacteria with nonresistant forms of the gene are unable to form properly shaped cell walls in the presence of penicillin, and are killed by the drug. Other genes might cause more active countermeasures, like pumping an antibiotic out of the cell or modifying the antibiotic to be less toxic to the cell. Browsing the UniProt database gives you a feel for the huge variety of genes available in nature that can make bacteria resistant to drugs.

THE DECOMPILER

Next, I needed the actual decompiler. That's where BLASTX (eventually updated to BLAST+) came in. BLASTX is a variant of BLAST, which stands for *Basic Local Alignment Search Tool.* First, I had this analysis program compute all possible translations of the *E. coli* DNA to protein sequences. Translating DNA results in six possible protein sequences: DNA can be read forward and backward (known as 5′→3′ and 3′→5′), and each direction has three possible frame positions. Then, I had the program check for patterns among the resulting amino acid sequences that matched the database of sequences known to provide drug resistance. (I could have also checked for other types of patterns, by typing something different into the database query.) The result was a sorted list of each known drug resistance protein, along with the region of the *E. coli* genome that best matches the protein.

The following is the BLASTX output for the penicillin example.

---

* Incidentally, I find it amusing that the sequence for PBP2 is shorter than, for example, my PGP public key block.

```
# BLASTX 2.2.24 [Aug-08-2010]

# Query: 43 87880
# Database: uniprot-drug-resistance-AND-organism-coli.fasta
# Fields: Query id, Subject id, % identity, alignment length,
mismatches, gap openings, q. start, q. end, s. start, s. end,
e-value, bit score
43 sp|P0AD65|PBP2_ECOLI 100.00 632 0 0 29076 30971 1 632 0.0 1281
43 sp|P0AD68|FTSI_ECOLI 25.08 650 458 21 29064 30926 6 574 2e-33 142
43 sp|P60752|MSBA_ECOLI 32.80 186 120 6 12144 12686 378 558 6e-17 87.0
43 sp|P60752|MSBA_ECOLI 27.78 216 148 5 77054 77677 361 566 8e-14 76.6
43 sp|P77265|MDLA_ECOLI 27.98 193 133 6 12141 12701 370 555 2e-10 65.5

--snip--
```

The `Fields` line describes what each column in the table shows. In the `% identity` column, you can see that the gene for PBP2_ECOLI has a 100 percent match inside the genome of O104:H4.

Answering Biological Questions with UNIX Shell Scripts

With this list, I could answer some interesting questions, like "How many of the known drug resistance genes are inside O104:H4?" Here's the one-liner program that my perlfriend wrote to answer that particular question:

```
cat uniprot_search_m9 | awk '{if ($3 == 100) { print;}}' | \
  cut -f2 |grep -v ^# | cut -f1 -d"_" | cut -f3 -d"|" | \
  sort | uniq | wc -l
```

The output from that script told us that 1,138 genes in O104:H4 were a 100 percent match against the database of 1,378 genes that can confer drug resistance. When we loosened the criteria to also list 99 percent matches, allowing for one or two mutations per gene, the list expanded to 1,224 out of 1,378. The "superbug" O104:H4 earned its title, having acquired roughly 90 percent of the known resistance genes!

I also wanted to answer the inverse question: which drug-resistance genes are most definitely not in O104:H4? By looking at the resistance genes missing from a superbug, we might be able to gather clues as to which treatments could be effective against the bug.

To rule out a drug-resistance gene, we crafted another search that would reveal which resistance genes in the database had less than a 70 percent match against the sequence of O104:H4. The 70 percent threshold was just an arbitrary number I picked; there's probably a rigorous standard that scientists and clinicians use.

Here is the list, as it appeared in my terminal:

```
A0SKI3 A2I604 A3RLX9 A3RLY0 A3RLY1 A5H8A5 B0FMU1 B1A3K9 B1LGD9
B3HN85 B3HN86 B3HP88B5AG18 B6ECG5 B7MM15 B7MUI1 B7NQ58 B7NQ59
B7TR24 BLR CML D2I9F6 D5D1U9 D5D1Z3 D5KLY6 D6JAN9 D7XST0 D7Z7R4
D7Z7W9 D7ZDQ3 D7ZDQ4 D8BAY2 D8BEX8 D8BEX9 DYR21 DYR22 DYR23
E0QC79 E0QC80 E0QE33 E0QF09 E0QF10 E0QYN4 E1J2I1 E1S2P1 E1S2P2
E1S382 E3PYR0 E3UI84 E3XPK9 E3XPQ2 E4P490 E5ZP70 E6A4R5 E6A4R6
E6ASX0 E6AT17 E6B2K3 E6BS59 E7JQV0 E7JQZ4 E7U5T3 E9U1P2 E9UGM7
E9VGQ2 E9VX03 E9Y7L7 O85667 Q05172 Q08JA7 Q0PH37 Q0T948 Q0T949
Q0TI28 Q1R2Q2 Q1R2Q3 Q3HNE8 Q4HG53 Q4HG54 Q4HGV8 Q4HGV9 Q4HH67
Q4U1X2 Q4U1X5 Q50JE7 Q51348 Q56QZ5 Q56QZ8 Q5DUC3 Q5UNL3 Q6PMN4
Q6RGG1 Q6RGG2 Q75WM3 Q79CI3 Q79D79 Q79DQ2 Q79DX9 Q79IE6 Q79JG0
Q7BNC7 Q83TT7 Q83ZP7 Q8G9W6 Q8G9W7 Q8GJ08 Q8VNN1 Q93MZ2 Q99399
Q9F0D9 Q9F0S4 Q9F7C0 Q9F8W2 Q9L798
```

You can plug any of these protein codes into the UniProt database and find out more about them. For example, BLR is beta-lactamase, an enzyme that causes resistance to beta-lactam antibiotics. UniProt describes it like this:

> Has an effect on the susceptibility to a number of antibiotics involved in peptidoglycan biosynthesis. Acts with beta lactams, D-cycloserine and bacitracin. Has no effect on the susceptibility to tetracycline, chloramphenicol, gentamicin, fosfomycin, vacomycin or quinolones. Might enhance drug exit by being part of multisubunit efflux pump. Might also be involved in cell wall biosynthesis.

Unfortunately, a cursory inspection revealed that most functions that O104:H4 lacked were just small, poorly understood fragments of machines involved in drug resistance. As a result, there was no clear candidate for a superbug killer in its genome.

More Questions Than Answers

The good news is that anyone can access the tools to analyze genomes, and some tools, such as `grep`, `awk`, and `sed`, are already familiar to computer engineers. The bad news is that while we can ask questions about the genome with these tools, we're still left with more questions than answers. For example, antibiotic resistance sounds like a good thing for the survival of bacteria, so why don't all bacteria have it? And how do bacteria go about acquiring (or losing) such genes?

The rise of antibiotic-resistant superbugs is a product of our love of antibiotics. As DNA in *E. coli* copies at a rate of about a dozen base pairs per second, shedding even a single unused gene can lend a meaningful advantage in an exponential growth race; after all, an *E. coli* population can double every 20 minutes in optimal conditions. As a result, there is selective pressure to shed genes that aren't necessary for survival. The genome of O104:H4 is 15 percent longer than that of a typical *E. coli* strain, which means that after seven generations, a typical *E. coli* strain would have twice the population of O104:H4. Within half a day under optimal, antibiotic-free growth conditions, a strain of *E. coli* unburdened with antibiotic resistance genes would have over 20 times the population of O104:H4. Thus, a bacterium that hangs on to its antibiotic resistance genes is like a sprinter wearing a bulletproof vest to a race. Likewise, one of the greatest natural threats to superbugs is a lean, fast-replicating common bug that can edge out the superbug by sheer numbers alone.

However, bacteriocidal and bacteriostatic antibiotics kill off or prohibit growth of nonresistant bugs, respectively, leaving only the resistant bugs to grow unhindered. Over time and with exposure to several types of antibiotics, it stands to reason that the resistant bug population would continue to selectively breed for multiple resistance genes, creating a superbug.

Still, I find it astonishing that resistant bugs seem to develop resistance genes so quickly. We're taught that evolution is a slow process, so it seems remarkable that bacteria can serendipitously evolve a suite of antibiotic resistance genes totaling hundreds of thousands of base pairs. New genes do in fact take a very long time to spontaneously arise (there are very few clearly documented cases of this, such as the Long-Term Evolution Experiment by Richard Lenski). Instead, most resistance genes are acquired from the environment through *horizontal gene transfer*.

Our environment is teeming with DNA fragments. The GitHub of biology is all around us, from the dirt to the sea to the air we breathe. Some DNA fragments code for useful traits; some are just junk. When a bacterium is under stress (like it is when exposed to antibiotics), it may start to take up random DNA fragments from the environment and manufacture proteins based off the code. If it's going to die anyway, it might as well, right? Most of the time, the incorporated DNA fragments are not helpful, but if one lucky bacterium picks up the necessary resistance gene from the environment, it can rapidly outcompete others in an antibiotic-laden environment.

Thus, while nonresistant strains of a bug will rapidly outnumber antibiotic-resistant strains, the tiny remaining population of resistant bugs (or perhaps even their lifeless bodies floating about in the environment) form a reservoir of genetic material that can be drafted in times of stress. And since the genetic code is interoperable across all species, resistance genes can even be acquired from unrelated organisms.

Discovering that the functions O104:H4 lacked were poorly understood was an interesting lesson in itself. Fiction popularizes the notion that knowing a DNA sequence is the same as knowing what diseases or traits an organism may have. But even though we know the sequences and general properties of many proteins, it's much harder to link proteins to a specific disease or trait. At some point, someone has to get their hands dirty and do biological experiments involving actual organisms to assign biological significance to a given protein family.

Pop culture references to DNA analysis are glibly unaware of this missing link in the process, which leads to overinflated expectations for genetic analysis, particularly in its utility for diagnosing and curing human disease and applications in eugenics. Let's take a closer look at some of those myths.

MYTHBUSTING PERSONALIZED GENOMICS

We're definitely living in The Future in a lot of ways. For instance, we have electric cars! But Hollywood reels from the '60s and '70s also predicted that I'd be using a flying car to get around town by now, not just an electric car on the ground. Of course, automotive technology isn't the only victim of Hollywood hype.

The potential impact of personalized genomics is greatly overstated in movies like *GATTACA*, which create a myth that your genome is like a crystal ball, and somehow your fate is predestined by your genetic programming. The perlfriend I mentioned earlier coauthored a paper in *Nature** examining 23andMe's direct-to-consumer (DTC) personal genomics offerings. Let's have a look at her paper, and let the mythbusting begin!

Myth: Having Your Genome Read Is Like Hex-Dumping the ROM of Your Computer

An inexpensive technique to look at parts of the genome is called *genotyping*. Here, a selective diff is done between your genome and a reference human genome; in other words, your

* P.C. Ng et al., "An Agenda for Personalized Medicine," *Nature* 461, no. 7265 (2009): 724–726, *http://www.nature.com/nature/journal/v461/n7265/full/461724a.html*.

genome is simply sampled in potentially interesting spots for single-point mutations called *single nucleotide polymorphisms* (SNPs, pronounced "snips"). The concept of genotyping naturally leads to two questions. First, how do you decide which SNPs are interesting enough to sample? And second, how do you know the reference genome is an accurate comparison point? This sets up two more busted myths.

Myth: We Know Which Mutations Predict Disease

Some mutations in the human genome simply correlate with disease; they are not proven to be predictive or causal. In truth, we really don't understand why many genetic diseases happen. For poorly understood diseases, all we can say is that people who have a particular disease tend to have a certain pattern of SNPs. It's important not to confuse causality with correlation.

Thus, while scientists can make predictions about diseases based on SNPs, most of those predictions are correlative, not causative (and weakly correlative, at that). As a result, a genotype should not be considered a crystal ball for predicting your disease future. Rather, it's closer to a Rorschach blot that you have to squint and stare at for a while before you can say what it means. For instance, in the paper my perlfriend wrote, she found that companies often didn't match up on their predictions for disease risk because they interpreted mutation meanings differently.

Myth: The Reference Genome Is an Accurate Reference

The word *reference* in *reference genome* should tip you off on a problem: it implies there are "reference people." Ultimately, just a handful of individuals were sequenced to create today's reference genome, and most of them are of European ancestry. As time goes on and more full-sequence genetic data is collected, the reference genome will be merged and massaged to present a more accurate picture of the overall human race, but for now, it's important to remember that a genotype study is a diff against a source repository of questionable universal validity.

For example, some SNPs have different frequencies in different populations. The base A might dominate in a European population, but at that same position in an African population, the base G could dominate. It's also important to remember that the reference genome has an aggregate error rate of about 1 error in 10,000 base pairs, although to be fair, the process of discovering a disease variant usually cleans up any errors in the reference genome for the relevant sequence regions.

It will be decades before we have a full understanding of what all the sequences in the human genome mean, and even then, they may not be truly predictive of disease risk or anything else about our health. Here lies perhaps the most important message, and a point I can't stress enough: in most situations, environment has more to do with who you are, what you will become, and what diseases you will have than your genes do. Any upside to personal genomics won't be due to crystal-ball predictions, but rather to the fact that knowing about their own genetic predispositions may encourage more people to make lifestyle changes that will help them stay healthy. If there's one thing I've learned from dating a preeminent bioinformaticist, it's that no matter your genetic makeup, most common diseases can be prevented or delayed with proper diet and exercise.

PATCHING A GENOME

So far in this chapter, I've given examples of sequencing and analyzing genomes. That's more or less the equivalent of being able to dump a program executable and analyze it in IDA. Oftentimes, after you analyze an executable, you'll want to patch it to do something new. Patching software is relatively straightforward and reliable: just fire up a hex editor and change the file. In the worst case, you might have to use a focused ion beam (FIB) to modify the individual wires of a mask ROM inside a chip.

But historically, the ability to patch a genome has been severely limited. Information in cells is stored at the molecular level, and changing a specific portion of a gene can be a painstaking process. Just as vacuum tubes and transistors came before the integrated circuit, zinc finger nucleases (ZFNs) and transcription activator-like effector nucleases (TALENs) enabled gene editing, but with significant caveats in efficiency, performance, and ultimately, cost. In 2012, the integrated circuit of gene editing was introduced: the CRISPR/Cas* system.

CRISPRs in Bacteria

CRISPR, short for *clustered regularly interspaced short palindromic repeat*, describes a particular RNA structure, while Cas are proteins that associate with CRISPRs. CRISPRs are, as far as biologists know, common only in bacteria and archaea (for example, fungi), and they're part of a devilishly clever system for immunity in simple organisms. Like humans, bacteria have immune systems that can be programmed through exposure to pathogens. When bacteria encounter a viral invader, they have proteins that can snip out short sequences of the viral DNA and archive the sequences as spacers in a CRISPR.

Labs that failed for months to edit a gene using TALENs switched to CRISPR/Cas and succeeded on the first try. They succeeded so quickly because the process just involves designing a short snippet of RNA that's inserted into a CRISPR, a simple exercise that can be done entirely on a computer or, I

---

* Addgene has an excellent white paper describing the system in great detail. I recommend checking it out if my cursory treatment here whets your appetite: *https://www.addgene.org/CRISPR/guide/*.

daresay, by hand. The RNA snippet itself can be fabricated in about a week for less than $50 using one of several service providers, replacing a significant amount of wet lab complexity with an informatics exercise.

Each CRISPR region is tagged by a leader sequence, immediately followed by the CRISPR proper. A CRISPR itself consists of a guide RNA (gRNA) or "spacer" sequence delimited by a well-defined DNA *direct repeat* sequence that is palindromic.

NOTE *The term* spacer *is used when discussing an immune system, while* guide RNA *is used when discussing genome editing. Calling a region of interest a spacer is confusing, but misnomers can happen with reverse engineering. I can't blame scientists for first noticing a pattern of repeating delimiters and calling the stuff between the delimiters "spacers." After all, physicists got the current flow convention backward and stuck with it. Who are we to judge?*

Palindromic typically means that a string is equivalent when simply reversed, like the word *racecar*. When biologists say a sequence is "palindromic," they mean the sequence is equivalent when first complemented (A→T, T→A, G→C, C→G) and then reversed. For instance, GAATTC is considered biologically palindromic, even though it is not lexically palindromic.

The CRISPR/Cas system was described shortly after the demise of Chumby, and at the time, I was interning at Dr. Swaine Chen's infectious diseases laboratory at the Genome Institute of Singapore. Among other things, I studied various strains of *E. coli* that induce urinary tract infection, under the guidance of Lu Ting Liow. While assisting an investigation into portions of phage virus DNA that found its way into *E. coli*, I was asked to write a script to identify palindromic and repeating sequences of DNA in the *E. coli* genome. My script showed that the genome was littered with the sequences; I figured the code had a bug and didn't think much of the result.

But perhaps some of the direct repeats I saw were portions of a CRISPR.

Let's look at a CRISPR from a strain of *E. coli* now. This is the CRISPR direct repeat sequence for *E. coli* O104:H4:

```
GAGTTCCCCGCGCCAGCGGGGATAAACCG
```

The bolded base pairs are the palindromic regions. When this DNA sequence is translated into RNA (so that T→U), the palindromic region can pair with itself, forming a hairpin or stem loop, as shown here.

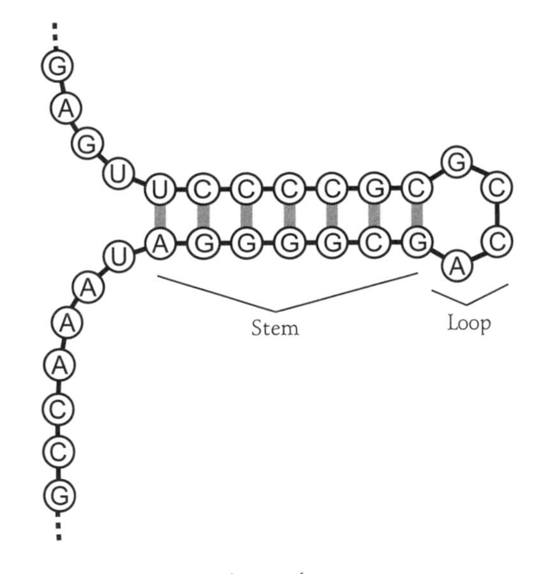

A stem loop

This shape hints at the significance of the repeated palindromic structures in a CRISPR: when translated into RNA, the sequence can fold onto itself, forming a *secondary structure*. It's important to remember that genes are not just lines of code; they are physical molecules whose overall shape significantly impacts their function. Biologists use a four-tier system for describing the physical structure that molecules like DNA, RNA, and proteins can take based on their source code. Primary structure is simply the sequence of monomers (bases or amino acids). *Secondary structure* refers to physical shapes that arise from the localized interactions of monomers, due to physical properties such as the spacing and number of hydrogen bonds between molecules, or the affinity of certain monomers for water. In RNA and DNA, that means structures like hairpin loops; in proteins, it means structures like spirals and sheets. Tertiary structure refers to the complex 3D shape of a molecule that arises from long-distance interactions between potentially remote portions of the primary sequence. Tertiary structure is particularly applicable to proteins, as some amino acids, such as cysteine, can cross-link with each other over longer distances. Quaternary structure refers to structures formed from the interaction of multiple molecules. A Cas9/RNA complex is an example of a quaternary structure. The final, chemically active and targeted molecule arises only when a Cas9 protein is merged with a gRNA, and the stem loop secondary structure of the gRNA is necessary for Cas9 to recognize it.

Determining Where to Cut a Gene

RNA derived from a CRISPR region through transcription is incorporated into a protein complex with other Cas proteins. Specific Cas proteins (such as Cas9) use the RNA as a search-and-destroy template: the Cas9/RNA complexes float around the cell, and when they find a DNA sequence that matches the RNA template, they selectively cut the DNA at the template site, effectively neutralizing the intruding virus. But you may have noticed a recursion problem: the Cas9/RNA complex should also cut up the CRISPR region in the host organism's genome, as that region also has the target pattern. This would effectively destroy the CRISPR region for future use.

To avoid destroying the CRISPR region, the Cas9/RNA complex targets the template DNA plus a short, defined three-to-five base pair sequence called a *proto-space adjacent motif*

(PAM). For example, the PAM for a popular Cas9 protein from *S. pyogenes* is [AGTC]GG when written in regular expression format; biologists use a different convention, NGG, to say the same thing. As long as the CRISPR archive doesn't include the PAM sequence, it won't be cut up by the complex.

The PAM requirement means there are some limitations on where you can cut a gene. It's a bit like targeting only hex strings that end in 0xC3 or searching for return-oriented programming (ROP) gadgets. Just as hackers searching for ROP gadgets look for short sequences of instructions that end in a RET opcode, bioinformaticists have to search for short sequences of DNA to edit that end in a PAM.

Despite these limitations, CRISPR/Cas has proven to be a versatile and reliable gene-editing tool. It has been adapted to both cut genes and paste in new sequences. Making a precise cut at an arbitrary location in DNA is the hardest step of inserting new DNA. But in conjunction with well-studied techniques like non-homologous end joining (NHEJ) or homology-directed repair (HDR), CRISPR/Cas can be used to insert modifications into a gene.

Implications for Engineering Humans

Even though CRISPR/Cas is a naturally occurring system found in bacteria and fungi, the universal genetic code means the system is binary-compatible with all species, including humans. Before this system was discovered, genes were largely read-only, especially in living organisms. CRISPR/Cas gives us a much more reliable and efficient tool to patch and repair genes, without necessarily disrupting the viability of the host organism. Biologists have managed to pack the necessary DNA for a CRISPR/Cas exploit into viruses, enabling them to sneak these gene-editing tools through the cell walls of live, complex organisms like mice, plants, and humans. The structure of a CRISPR also allows scientists to perform multiple edits in a single experiment, expanding the experimental and therapeutic versatility of the technique.

This technology has already been validated on human cells, even human embryos, and the implications are simply mind-boggling. Regardless of ethical standards set by the scientific and legal communities in your country of residence, I think the promise of custom-designed children, free of genetic diseases that once plagued parents, is too strong a temptation. Even if most countries banned such a practice, I feel it's inevitable that someone, somewhere, perhaps funded by a wealthy billionaire unable to have viable children of their own, will start tinkering with custom-engineered humans. If the results are positive, it will likely change the course of humanity more profoundly than Moore's law. And that's if a mechanism called *gene drive* doesn't get there first.

Hacking Evolution with Gene Drive

Gene drive rewrites the rules of sexual reproduction and, consequently, evolution in a way previously unseen in nature. You might know that you have two copies of every gene: one from your mother and one from your father. Each copy is an *allele*. If the alleles match, you're said to be *homozygous* for that gene. If the alleles are different, you're *heterozygous* for it. Normally, which allele a child gets from each parent is a coin toss, and the fitness of a child in a given environment is the primary deciding factor for passing a set of alleles on to a new generation.

Gene drive eliminates this coin toss. Environmental selection is short-circuited, allowing genes with potentially negative side effects to propagate rapidly in a population. This exploit is made possible by outfitting the desired allele with a CRISPR/Cas-assisted gene-editing mechanism that targets and converts a heterozygous allele into a homozygous allele. For example, if a mother has a gene outfitted with a CRISPR/Cas-assisted gene drive mechanism, it doesn't matter what the father's genes are. Inside the child, the mother's copy will express the CRISPR/Cas editing mechanisms, seeking out the father's copy and editing it to be the same as the mother's.

In terms of disruptive power, if CRISPR/Cas is the `rm` command, then gene drive is like calling `rm -r *` instead.

This has a profound effect on natural selection. Forget survival of the fittest; changes no longer have to strictly benefit an organism's fitness to spread through the population. Furthermore, gene-driven changes can sweep through a natural population at an exponential rate (much faster than typical mutations) because they don't rely on coin tosses and natural selection to amplify a mutation.

On the upside, gene drive could be used to force good changes into the world, like malaria-free mosquitoes. On the downside, this new mechanism, previously unseen in nature, could wreck havoc on evolution and the ecosystem. Although our changes could be well engineered and well intentioned, nature likes to shake things up through mutations, spontaneous rearrangements, and horizontal gene transfer. If a gene-driven organism were to pick up extra genes in the payload region, the outcome could be unpredictable.

For instance, malaria-free mosquitoes would benefit humans, but mosquitoes also play a large role in the Earth's ecosystem as a food source for fish and birds. If modified mosquitoes failed to thrive and occupy their ecological niche, there could be a domino effect that hurts other species. This could all happen on a timescale so short that we may not be able to reverse it if we tried. Furthermore, organisms like mosquitoes don't recognize geopolitical boundaries. Thus, banning gene drive in most of the world doesn't make anyone safe from its potential consequences. If just one well-engineered organism makes it into the wild, everyone has to deal with it.

Perhaps it's no mistake that CRISPR/Cas has been found only in bacteria and archaea—organisms that are known to reproduce asexually. Perhaps the ability to short-circuit the fitness requirement in sexual reproduction rapidly degrades the overall fitness of any germ line carrying a CRISPR/Cas mutation so that the line goes extinct before it can take over a population. After all, any accidental genes or spontaneous mutation that finds its way into a CRISPR/Cas payload would also sweep through the population as quickly as the initial drive.

The question, then, is how long does it take for this degradation and extinction to happen? The example of eradicating malaria vectors would have a very different outcome if the modified mosquitoes went extinct within a few years versus several millennia.

CLOSING THOUGHTS

Clearly, there are a lot of unanswered questions on the frontier of biological engineering, and it's all happening right now. Whether good or bad, the outcome of today's experiments will probably affect humanity as profoundly as Moore's law and the internet. Electronic technology reshaped the way we think and communicate, and biotech will reshape our bodies and our environment. The big difference is that in biotech, we haven't developed the ability to do backups, but we are developing technology with the potential power of the `rm -r *` command.

Personally, I'm optimistic; I think these technologies can and will be used to improve our lives. But for that to happen, we need society to understand the issues at stake and have a vigorous and open debate. Even if these biological techniques have scary implications for our health and safety, failing to disclose and discuss vulnerabilities just invites zero-days. And who wants to wake up one morning infected with crippling malware and no viable patch?

Hardware breakthroughs have changed our lives as we know it, but Moore's law is slowing down, and DNA sequencing has outpaced it. Who knows what new world will be created by advancements in biotech? And just as society benefits from the responsible disclosure and sharing of vulnerabilities and exploits, engaging in scientific discourse is more constructive than attempting to censor it. Perhaps the experience and perspectives gained in maturing the hardware industry over the past 50 years from pocket calculators into pocket supercomputers can help guide biotech to a similarly positive outcome.

11. selected interviews

I've done several interviews over the years, and this chapter compiles a couple that I thought you might enjoy. The first interview was originally published by the *China Software Developer Network (CSDN)*, which describes itself as a "programmer magazine." At the end, you'll find a story from the *Blueprint*, a collection of interviews with founders and innovators in hardware.

ANDREW "BUNNIE" HUANG: HARDWARE HACKER (CSDN)

This interview originally appeared in *CSDN* in Chinese in 2013, and the magazine kindly allowed me to publish an English translation on my blog. In the first section, I discussed my thoughts on the maker movement, which was relatively new at the time, and my experience with making hardware products. The second section was more about hardware hacking and what I feel it means to have a hacker spirit. You can find the original Chinese-language version at *http://www.csdn.net/article/2013-07-03/2816095*.

About Open Hardware and the Maker Movement

The maker and open hardware movements have attracted a lot of attention. Chris Anderson wrote a book called Makers, and Paul Graham called this time the "Hardware Renaissance." How do you think this movement will affect ordinary people, developers, and our IT industry?

This movement, as it may be, is more a symptom than a cause, in my opinion. First, let's review how we got to this point.

In 1960, there was only hardware, and it was all open. When you bought a transistor radio, it had a schematic printed in the back. If the radio broke, you had to fix it yourself. It was popular to buy kits to make your own radios.

Between 1980 and 1990, the personal computer revolution began. Computers started to become powerful enough to run software that was interesting and enabling.

From 1990 to 2005, Moore's law drove computers to be twice as fast and have twice as much memory every 1.5 to 2 years. Only software mattered, because unless you could afford to fab a chip in the latest technology, making hardware wasn't worth it. By the time you got the components together, a new chip would make your design look slow. Optimizing software also mattered less than features, convenience, and creativity. Users could just buy a faster computer and run old software faster. "Making" fell out of fashion because there was no time for it: you had to ship code or die.

From 2005 to 2010, computers didn't get much faster in terms of clock speed, but they got smaller. Smartphones were born. Everything became an app, and everything is still becoming more connected.

From about 2010 to now, Moore's law has been slowing down. This slowdown is rippling through the innovation chain. PCs aren't getting faster, better, or cheaper in a meaningful way. We buy new PCs just to replace broken ones, not because the latest model is so much better. It's too early to tell, but smartphones may also be solidifying as a platform: the iPhone 5 is quite similar to the iPhone 4, and Samsung phones also look pretty similar across revisions.

The question, then, is how to innovate? How can you create market differentiation? With Moore's law slowing down, it's possible to innovate in hardware and not have your innovation look slow because a new chip came out. You have steady platforms (PCs, smartphones, tablets) that you can target your hardware ideas toward. You don't have to fab chips just to have an advantage. Everyone is now sifting through technology's past, looking for niches that were overlooked. Even an outdated smartphone motherboard looks amazing when you put it in a quadcopter, satellite, HVAC system, automobile, energy monitoring system, health monitoring system, and so on.

Furthermore, as humans, we fundamentally feel differently toward physical things and virtual things. Apps are wonderful, but human homes are more than a smartphone, a food tray, a bed, and a toilet. People still surround themselves with knickknacks, photos of friends, and physical gifts from special occasions. I don't think there will ever be a time when a virtual teddy bear app will displace a physical teddy bear for cuddling at night.

As a result, there will always be a place for people to make hardware that fills this need for tangible goods. This hardware will merge more technology and run more software, but in the end, there is a space for makers and hardware startups, and that space is just getting bigger now that hardware technology is stabilizing.

Arduino and Raspberry Pi seem to reduce the threshold for designing hardware. How do you think this will affect the hardware industry? Do you think these platforms will progress the industry by leaps and bounds? If not, what does it take to make a really innovative hardware product?

Arduino and Raspberry Pi serve specific market niches.

Arduino's key contribution is reducing computation to an easy-to-use physical form. It was made first and foremost by designers and artists, and less so by technologists. This unique perspective on technology is very powerful because people who aren't programmers or hardware designers want to access hardware technology, too. Some very moving, deep interactive art pieces have been made using the Arduino, allowing hardware to transform menial control applications into artwork that changes your mood or makes you think about life differently. I think Arduino is just the first step toward taking the "tech" out of technology and letting everyday people not just use technology but create with it. There will be other platforms, for sure.

Raspberry Pi is a very inexpensive embedded hardware reference module, and I think other platforms will follow in its footsteps. It's cheap enough that for many applications, you can use the Raspberry Pi as is and gain no net cost advantage by designing and building your own hardware. For hardware professionals, the nice thing about this platform is that instead of buying a reference design and then having to spin your own board, you can just buy the Raspberry Pi and ship it in your product. For people who have relatively low-volume products, this makes sense.

I see an ongoing trend toward product design becoming more feasible at low volumes. There's still a market for million-unit blockbuster devices like smartphones and coffeemakers, but eventually, there will also be a market for devices that only have a production run of 1,000 to 10,000 units, but with a much higher margin. These small-run products will be developed and sold by teams of just one or two people so that the profit will still be a good living for the individuals. The key to the success for these products is that they are highly customized and help solve a specific problem for a small group of users who are willing to pay more for the solution.

When new concepts or technologies first appear, they always generate optimistic discussion, but most of them will really affect our lives only after a long period of development. When discussing the maker and open hardware movements, are we too optimistic? Does the average person have common misunderstandings about this field?

Yes, it does take a long time for technology to really change our lives.

The maker movement, I think, is less about developing products and more about developing people. It's about helping people realize that because technology is man-made, every person has the power to control it with a little knowledge. There is no magic in technology. You could also say that anyone can be a magician with a little training.

Open hardware is more of a philosophy. The success or failure of a product is largely disconnected from whether the hardware is open or closed. Closing hardware doesn't stop people from cloning or copying, and opening hardware doesn't mean that bad ideas will be copied simply because they are open. Unlike software, hardware requires a supply chain, distribution, and a network of relationships to build it at a low cost. That overhead means being open or closed is only a small part of the equation, and the question of whether to open or close a project revolves around how much you want to involve end users or third parties to modify or interoperate with your product.

Looking at the future of open source hardware, do you think it will be analogous to the open source software industry, where many commercial companies also support open source software? What are the differences between them?

I don't think they're quite analogous. In software, the cost to copy, modify, and distribute is basically zero. I can clone a copy of the Linux source repository, run the `make` command, and have the same high-quality kernel running on my desktop that runs on top-end servers and supercomputers.

But copying hardware has a real cost: the parts, the factories, and the skilled workers used to build them; the quality control procedures; and the manufacturing process are all important factors in the final product's cost, look, feel, and performance. Simply giving someone a copy of my schematics and drawings doesn't mean they can make my exact product. Even injection molding has art to it. If I give the same CAD drawing to two tooling makers, the outcome could be very different depending on where the mold maker decides to place the gates, the ejector pins, the cooling for the mold, the mold cycle time, temperature, and so on.

And then you have to think about the distribution channel, reverse logistics, financing, and so on. Even as the world becomes more efficient at logistics, you'll never be able to buy a TV as easily as you can download the movies that you'd watch on that TV.

What kind of business model do you think is ideal for an open source hardware company? Could you give an example?

One of my key theories behind open source hardware is that regardless of the license, hardware is essentially open, at least at the level of schematics and PCB layout. For a relatively small amount of money, you can pay a service to extract the details required to copy a PCB design. Therefore, you can assume that once you ship hardware, it can be copied. If you accept this assumption, then it follows that not releasing schematics and PCB layouts won't stop people from copying your goods. If someone wants to copy a piece of hardware, they will, whether you share your design files or not.

But sharing design files does make a difference to a separate and important group of people. There are other businesses and individual innovators who could use your design files to design accessories, upgrades, or third-party enhancements that rely upon your product. In that case, sharing your design files improves your opportunity for new business relationships, which makes doing so (with an open source hardware license to reserve a few basic rights and protections) a practical suggestion.

Clearly, some hardware strategies aren't compatible with open source. If your sole value to the consumer is your ability to make stand-alone hardware, and you have no strategic advantage in terms of cost, then you'd want to keep your plans secret to delay low-cost copies for as long as possible.

But the most innovative products today aren't just pieces of hardware. They also involve software and services. Open hardware business models work better in such hybrid products. In many cases, consumers are willing to pay annually (think in terms of subscriptions, advertising, upsells, accessories, royalties, or upgrades) for many products. In fact, it's most profitable to just collect these fees and not involve yourself in the hardware manufacturing portion. Controlling access to an ongoing service is also much easier than controlling the plans for a piece of hardware.

Thus, if you couple a profitable online service with your hardware, open hardware makes a lot of sense. Letting other people copy the hardware, sell it, and add more users to your online service simply means you get more revenue without more risk.

You come to China often and know a lot about this country. China's software technology is not advanced. Do you think that being the world factory center will help China improve its overall level of technology? How can this country change from just a manufacturing center to a place focused on design, research, and development? What is China missing?

I wouldn't say I know much about China. I know a little about one small corner of China in one specific area—hardware manufacturing. If there's one thing I do know, however, it's that China is a very big country with many different kinds of people and a long history that I am only beginning to understand. However, I've lived through almost the entire history of high technology, so I can comment on the relationship between high technology and people, from which I can derive some perspective about China.

First, every country that is a technology powerhouse today started with manufacturing. The United States started as colonies of Britain, mining ores, trapping furs, and farming cotton and tobacco. Over time, the United States had steel mills and linen production. The United States didn't really start to develop original technology until the early 1900s, and that process didn't take off until the mid 1900s.

Japan developed similarly. It started in manufacturing, copying many US-made goods. In fact, if you believe the historical accounts, the first cars and radios made in Japan were not great. It took the United States and Japan decades to go from manufacturing-based economies to service-based economies.

Compare that to China, where the electronics manufacturing industry started maybe 20 years ago, at most, and China is just turning the corner from being a manufacturing-oriented economy to one that can do more design and software technology. I believe this is a natural series of events. Some portion of entry-level workers will eventually become technicians, then some technicians will become designers, and finally, some designers will become successful entrepreneurs.

In concrete numbers, if you have 10 million factory workers, maybe 1 percent, or 100,000 workers, will learn enough to become technicians after a few years. After a few years of technician work, maybe 1 percent will gain enough skill to become original designers, giving 1,000 designers. These experienced, grassroots designers would become the core of an entrepreneurial economy, and from there, the economy could begin to transform.

Over the course of a decade or two, a thousand companies would eventually be distilled to just a handful of global brand companies. I believe China is currently going through this final phase. A lot of people in Shenzhen have the experience of manufacturing, the wisdom to do design, and the ability to apply their talent to innovation and original product design. The next decade will be an exciting one for China's technology industry, if the current policies on economic and intellectual development stay roughly on course.

This pattern applies primarily to hardware or hardware-dominated products. Software products have a similar pattern, but I believe there are unique cultural aspects that give the West an advantage in software design. In hardware, if a process is not efficient or is producing low yield, you can easily identify the root cause and produce direct physical evidence of the problem. Hardware problems, in essence, are indisputable.

In software, if code is not efficient or it's poorly written, it's very hard to identify the exact problem that causes it. You can see evidence of programs crashing or running slowly, but there's no broken wire or missing screw you can hold up to show everyone why the software is broken. Instead, developers have to review complex designs, consider many opinions, and ultimately, identify a problem that comes down to nothing more than one individual's bad decision. All software APIs are simply constructs of human opinions.

Asian cultures have a strong focus on *guanxi*, reputation, and respect for the elders. The West tends to be more rebellious and willing to accept outsiders as champions, and they have less respect for the advice of elders. As a result, I think it's very culturally difficult in an Asian context to discuss code quality and architectural decisions. The field of software itself is only 30 years old, and older, more experienced engineers are also the most out of date in terms of methodology and knowledge. In fact, the young engineers often have the best ideas. But if it's culturally difficult for young engineers to challenge the decisions of elder engineers, you end up with poorly architected code and no hope to be competitive.

Overcoming these obstacles is possible, but enforcing the correct incentives and culture would require a very strong management philosophy. The workers should be rewarded fairly for making correct decisions, and there can be no favorites based upon friendship, relationship, or seniority. Senior engineers and managers must see a real financial reward for accepting their mistakes, instead of saving face by forcing junior engineers to code patches around bad high-level decisions. US companies usually achieve this alignment by sharing equity in a company among the engineers so that the big payout only comes if the company as a whole survives, regardless of an individual's ego.

What do you think the relationship between individual makers and commercial companies will be in the future? And as individual makers may compete not only with commercial companies but also with other makers in the future, what factors are critical to a product's success?

As minimum order quantities decrease and innovation gets closer to the edge, I think commercial companies will see more competition from makers, especially as the logistics industry transforms itself into an API that can plug directly into websites. At the end of the day, the most critical factor to success will still be how much value consumers perceive from a product. This is related to superior features and good product quality, but the presentation to the consumer and how clearly the benefits are explained are important, too.

As a result, any product will need to be visually appealing, be easy to use, and come with marketing material that clearly explains the benefits of using it. Those elements are often challenging for individual makers who are good at making products that are valuable technically but have less talent for sales and marketing. Makers who can master both facets will have an edge.

About Hardware Hackers

You've participated in the development process of many products, but what is your personal goal?

I would like to make people happy by building things that improve their life in some way. The greatest pleasure is to see someone enjoying something I made, and knowing I've improved that person's life in some small way. Sometimes, the product is solving a big problem for its users; other times, the product is more whimsical, and the user's happiness comes from fun or beauty. But either way, knowing I'm helping another person by making something is important to me. I've learned that money beyond a certain level doesn't make me any happier. This makes me difficult to work with, because it's hard for people to just hire me by offering a lot of money. Instead, they need to convince me that the activity will somehow also make people happy.

Another important goal for me is to just understand how the world works. I have a natural curiosity, and I want to learn and understand all kinds of things. The universe has a lot of patterns to it, and sometimes, you'll find seemingly unrelated pieces fitting together like magic. Discovering these links and seeing the world fit together like a big jigsaw puzzle is profound and satisfying.

Failure tends to give people more experience. Could you talk about the not-so-successful projects you have participated in, or if you've ever seen other failed projects that inspired you?

My life is a story of failures. The only thing I have done repeatedly and reliably is fail. But I have two rules when handling failure:

1. Don't give up.
2. Don't make the same mistake twice.

If you follow these rules, eventually, you'll find success after many failures. That said, I do have an interview that focuses on one of my recent failures. You can read it at *http://makezine.com/2012/04/30/makes-exclusive-interview-with-andrew-bunnie-huang-the-end-of-chumby-new-adventures/*.*

* This interview is excerpted in Chapter 6.

Your book,* Hacking the Xbox, *has been published for 10 years. For people who want to learn reverse engineering or become a hardware hacker today, how do these experiences and skills still apply?

I'd like to think the core principles covered in the book are still relevant today. The Xbox was simply an example I used to show how to do things. The approach and the techniques are applicable to a broad range of problems.

For the Chinese audience, I have found mobile phone repair manuals to be quite interesting to read, even though I can't read Chinese well. Their descriptions on the theory of electronics are not always completely accurate, but practically speaking, they're good enough, and they provide a quick way to get started while learning immediately useful skills in repairing phones.

There's also a Chinese magazine, called 无线电 (something like *Radio Electronics* in English), which I have found to be quite good. If you get started building the projects in there, I think you will learn very quickly.

The Xbox One has more stringent restrictions for users. What do you think about this? Are you interested in exploring this black box and upgrading your book?

I haven't done much work on video game consoles in a while; there's a whole new generation of console hackers who are excited to explore them, and I'm happy for that. As for the Xbox One's security, I'm sure it is one of the most secure systems built. Microsoft did a very good job on the Xbox 360, and the Xbox One security team members I know personally have a very solid understanding of the principles needed to build secure hardware. It should be very hard to crack.

That said, I'm glad I have no desire to buy or use one. I think I would become very frustrated with their use policies and restrictions very quickly.

There's a lot of controversy over whether electronic devices should have a lock to prevent user rooting. What do you think about this? Is there a contradiction between ensuring user safety and giving users complete control of their devices?

I believe users should own their hardware, and owning something means having the right to modify it and having root access rights. If a company is concerned about users being unsafe, then it's easy enough to allow users to opt out by signing an electronic waiver to give up their support and warranty rights in order to gain complete access to their own machines. Most people who can root their machines are already smarter than the phone support they would be calling inside the company, so they shouldn't have problems.

The laws have changed to make some rooting activities illegal, even on hardware that you bought and own. I think this reduction in our natural rights of ownership is dangerous and can put consumers in unfair situations. This also discourages consumers from exploring and learning more about the technologies they've become so dependent upon.

As hardware systems become more integrated, do you think hardware hacking is getting more and more difficult, or do you worry about hardware hackers becoming extinct? If so, how could we change this situation?

Hardware system integration has been increasing for a long time. The TX-0 just used transistors, the Apple II used TTL ICs, PCs use controller chipsets, and mobile phones have just a single System-on-Chip. Increasing integration does make some parts harder to hack, but there are always opportunities at the system integration level.

In other words, I still think there is art in hardware, but the level at which hardware hackers have to work gets higher every day, and that's a good thing. It means hacks are getting more powerful with time as well.

Hacking the Xbox *is dedicated to Aaron Swartz. Could you talk about why you think the hacker spirit is important today?*

The hacker spirit is the ultimate expression of human problem-solving ability. It's about the ability to see the world for what it is, and not the constructs and conventions that society puts in place. For instance, a brick is not just used to make buildings; it can be a doorstop, a weapon, a paperweight, a heating ballast, or it can be ground up and used for soil. Hackers question convention through the lens of doing what's most practical and correct for the situation at hand. Sometimes their methods aren't always harmonious, as hackers often prioritize doing the right thing over being nice or playing by the rules.

I find the more difficult situations become, the more pervasive and stronger the hacker spirit becomes among common people. I see evidence of this around the world. This spirit is linked to the human will to survive and to thrive. I think it's important for a society to cultivate and tolerate the hacker spirit. Not everyone has it, but the few who do help make society more resilient and survivable in hard times.

Do you have other words you would like to share with Chinese readers?

I was reading some comments on a Chinese web forum and was surprised that many Chinese regard the term *shanzhai* as a negative term. As an outsider, I feel that the shanzhai have done a lot of very interesting and useful innovation.

In English, we have a similar problem. The term *hacker* in English started as a good term but over time became associated with many kinds of negative acts. The term *maker* was coined to distinguish between the positive and negative aspects of hackers, but I still call myself a hacker because I still adhere to the traditional definition of the word.

It may be easier to explain the innovation happening in China if a similar linguistic bifurcation could happen in Chinese. I recently proposed referring to the innovative, open aspects of what the shanzhai do, like their method of sharing design files, as *gongkai* (公开). Significantly, I feel the term 开放 (*kai fang*, which means to lay open or to open to the public) as used in 开放源代码 (*kai fang yuan dai ma*, which means open source software) doesn't quite apply. It refers to a specific Western-centric legal aspect of being open, which is not applicable to the methods engaged in the Chinese ecosystem.

NOTE *Incidentally,* kai fang *also means to bloom, so it sounds poetic in Chinese.* Gongkai*, on the other hand, just means public or overt—whether you like it or not. Its meaning is not as poetic or optimistic as* kai fang.

The fact that China has found its own way to share IP, unique from the Western system, doesn't mean that the Chinese system is bad. It's actually quite interesting, and I'm very curious to see where it goes. Since I see positive value in some of the methods that the shanzhai use, I'd propose using the more positive, generic term *gongkai* to describe the style of IP sharing commonly used in China, but I would stop short of formally associating it with the strict definition of open source.

But then again, who am I to say? I'm not a native Chinese speaker, and maybe there is a much better way to address the situation.

THE BLUEPRINT TALKS TO ANDREW HUANG

The *Blueprint* publishes stories about founders in the hardware space, and this interview focuses on, as the writer put it, my "personal journey." I discuss what got me into hardware as a kid, what projects I was working on when I gave the interview, and pitfalls that hardware startups should keep an eye out for. The original interview, which includes some photos of my projects and answers to a few other interesting questions that didn't appear in the interview proper, is at *https://theblueprint.com/stories/andrew-huang/*.

How would you describe your first encounters with hardware?

My dad bought an Apple II clone when I was eight years old, and that sparked my interest in hardware. The clone came without a case, leaving all of the electronics exposed. I could see the electronics, and I wanted to fiddle with them. My dad didn't want me to touch the computer because I might break it, but when he wasn't home, I'd still fiddle with the electronics. I broke it several times because the chips were in sockets. Even though my dad told me not to, I just wanted to see what happened when you put the chips in backward. I learned very early on that putting chips in backward is a bad thing!

The great thing is that the Apple II came with a cool set of schematics and source code. I was the weird kid in elementary school who carried around an Apple II reference manual. On the playground, I'd just pull up the schematic and stare at it because it was so fascinating. I didn't understand what I was looking at, but I had some inkling about the connection between lines on the schematic and wires on the board. Over time, I learned to map the schematic's symbols to the computer functions bit-by-bit, and it all started coming together.

By junior high or high school, I was able to build my own plug-in cards for the computer, and I built a little speech synthesizer. That's what you do when you grow up among cornfields in Michigan and kids don't want to play with you because you look strange and you are the only Chinese kid.

How did your early experiences affect your decision to go into the hardware industry?

I just kept learning more from there. When I went to MIT, I flipped a coin, and instead of going into biology, I went into electronics. I got a degree, eventually went into industry, hated that, and then went back for my PhD because I wanted to hide in my shell a little more. After getting my PhD, I participated in a bunch of startups that all failed. I never had a successful startup, but I learned a lot from failure.

I did some silicon chip design and reverse engineering before I did manufacturing. For many years, I wanted to do the biggest, baddest, hardest project I could do, which meant working for a pure tech startup. With something like that, you're way in the future and basically by the time the technology works and goes onto the market, the patents have expired. There is no capital monetization, you work really hard, and the product is really obscure. As a result, I never had anything ship in volume. That was the most frustrating part: to put my life into something and never have it see the light of day.

What lessons did you learn while working on chumby?

I got tired of working for a pure tech company and decided it was time to join a company that could monetize a business idea quickly. When I joined Chumby, I wanted to do open hardware and manufacturing, and I started logging experience in both. I worked on the first chumby and then multiple generations after that from 2005 to 2010.

When I started, I had never mass-produced a product or done mechanical design. I didn't even know what injection molding was. But I had the privilege of sitting with other engineers at PCH, and I would just get on the factory floor, see what they were doing, and learn about it. By the time I was through with Chumby, I was able to use SolidWorks to design my own cases and make injection-molded cases from scratch.

It was a very educational experience. I learned to do test plans, manufacturing, sourcing, and other skills you just have to pick up along the way. When Chumby went under, I was living in Singapore, where I had attempted to open a field office. I stayed behind to wind down the office, give it a clean shutdown, and make sure everyone got jobs elsewhere. After everything was taken care of, I decided to be unemployed for one year; the first thing I did was design a radiation sensor for Japan after the terrible earthquake and tsunami on March 11, 2011.

Then I started thinking about what my next project would be. I did a series of projects like reverse engineering SD cards, and I met Jie Qi, who I helped to produce circuit stickers under the Chibitronics brand.

One of the guys working with me in Singapore was Sean Cross, and we were sitting around asking what we should build. We decided to build something we could use because when I was at Chumby, I built things for other people rather than myself. I use a laptop every day, and we needed a development platform, so we built a laptop that we would actually use. We're now doing a crowdfunding campaign around that product.

How would you describe your process of going from a prototype to manufacturing it?

There's actually a lot of art in designing things to be easy to make. One great approach to this is to be fully responsible for your own supply chain. I don't like to have a supply chain manager and a manufacturing manager. I want to make something myself. I insist on doing all of the testing myself. I insist on handling the manufacturing issues myself because, from a design standpoint, doing so forces you to think, "Can I build that? If I gloss over this bit of detail, I might pay dearly for that later."

From the very beginning when you start designing, I think about how to make something manufacturable. What manufacturing process should I use? How do I make sure I can source all of these components? When I actually get to the manufacturing time, I've made all the decisions because I'm the one who has to pay the price at the end of the day.

What do people most overlook when they are designing?

There are a lot of aspects you could forget. The two that come to mind first are the ability to source the materials and the yield. For example, the instructions for a cool project in *Make:* magazine often tell you to go find an obscure or out-of-date object, like a motor from a 1980s VHS player. In theory, that would be great because many people have this cheap item in their garage. But all of a sudden, everyone is going to eBay trying to find the same part, and it's not sourceable.

On the yield side, a lot of people won't run the numbers in terms of what it means to be yielding. Every step of the manufacturing process has some fallout. If every step is about 99 percent yield and you take 10 steps like that, your yield will be about 90 percent. People essentially build the Leaning Tower of Pisa into their project, and at the end of the day the problems compound, preventing delivery. It's crucial to build a system that is robust and reworkable so that every step can be coupled with another step to minimize yield fallouts. Otherwise, you'll throw away a lot of money.

How would you describe how things have changed in the perception of hardware since you got involved in manufacturing?

It's weird. Right around the time I was working on the Xbox in 2001, hardware was probably at the rock bottom. During the dot-com boom, working on Web 2.0 was really super-hot, and if you did something with Amazon or XML, it was cool. Soldering was a low-value thing that happened somewhere else.

But I was that weird guy who knew how to solder in a lab, so people would come to me with broken things and I'd fix them. I just stuck with it because that's what I do, and I love doing it. One reason the Xbox's security was relatively easy to break was because of the assumption that hardware was hard and soldering was difficult. But if you know how to solder, breaking the security is very easy. I did it on a grad school budget for about $150. I gave some talks at conferences after the Xbox hacking, basically telling people that hardware is not hard, that there's no magic behind it. I showed people that the "magic" was actually pretty simple manufacturing techniques.

Then Kickstarter came. Money started going into a system where it hadn't before because VCs wouldn't touch hardware. They thought hardware was a retail chasm where all this money had to be paid up front, then basically the startups all die, and investors don't get returns.

All of a sudden, these cool companies began raking in a million dollars in Kickstarter as their seed round and eventually delivering on their products enough of the time. There's nothing like money to get the interest of the guys in Silicon Valley. Since then, hardware perception has changed radically. People are starting to get into hardware more and more. The problem is that a lot of people think they have to add hardware to products now, yet have no idea how.

Another problem is an increasing number of scams on Kickstarter, where there are all these hardware bits and pieces, and backers can't tell what's real or what's fake. I know the industry definitely feels like a bubble already; I can sense the bubble growing now.

I think maybe I liked it better when nobody knew about hardware because at least I didn't have to worry about competing with fraudsters.

How have you approached finding your own factories?

If you're a startup and the only value you can bring to a factory is money, then you're basically worthless. Startups don't have any money, and if you have money, it's finite. All factories know this.

A lot of startups want to go to somewhere like Foxconn, but Foxconn has a ton of people and capability. They don't need your help. But they do need your money, and you don't have a lot of it. If you try to engage with the really hip factories, you'll deplete your cash very quickly and won't be able to launch.

I look for factories that are missing certain capabilities, so I can give them more value than money. When I come in with my product, I help train the staff to build my product. The factories see value in that training, and I get to that point where I'm building a relationship by trading more than money.

What's the challenge for online hardware startups when they get to the retail phase?

In the world of the internet, where everything is automated, it seems like you could solve any problem with technology. But retail is all about the salesperson meeting buyers face-to-face, doing demonstrations, and going to the Walmart or Target headquarters to actually develop relationships and cut deals. It feels like an older system, and a lot of people don't expect that because they're doing business with Kickstarter.

The problem is that people want to physically see and touch and feel a product before they spend a couple hundred dollars on it. Best Buy is becoming a showroom for Amazon, but offering the product in-store is really valuable. There is probably room for some disruption (perhaps you can convince credible reviewers to try your hardware and describe it to other people), but at the end of the day, retail presence is needed to sell hardware effectively.

Margins are much fatter online, so companies that start a business online from the beginning tend to underprice their products. Then, when they get to retail, they can't survive.

What are some of the most common questions that hardware entrepreneurs ask you?

The questions teams tend to ask usually center on weaknesses in their team composition. Some teams have super-hotshot electrical engineers, but they have no mechanical engineering background. These teams have a bunch of "mech-y" questions. Some teams have no electrical engineers, and then the big question is how to create a hardware startup with no one who can design electronics.

Hardware startup teams generally tend to be technical, so they're often weak on marketing and business. Some do have business guys involved early on who can map it all out and get a strategy in place, but a lot of teams have great tech ideas without realizing they're missing crucial aspects to their strategy.

At that point, I get them to tell me what they're doing, and I give feedback. It's almost not what teams ask, but rather what they forget to ask, that they need the most help with.

What do you think is missing from startups that will be necessary for the ongoing support of the hardware ecosystem?

There is a huge mismatch between the way manufacturing has been done and the way it needs to be done to match these more agile, lean, and honestly, less experienced companies. But I don't think it's an impassable chasm.

The original design manufacturers (ODMs) who have factories and resources need to raise their level of service. People expect ODMs to be able to answer a lot of questions. There are unreasonable expectations between startups and ODMs because ODMs can offer absolutely zero insight into costing down your product. People get upset because they just don't see that conflict of interest.

A lot of people think that building a product in China means the cost of parts gets magically cheaper. They don't understand. A factory is not a designer; its job is to ensure that your design works and is built to specification. If you specify an expensive part, and the factory substitutes a cheaper version, who gets the blame when the product doesn't work as well? Furthermore, the factory makes its money as a percentage margin over the bill of materials. Thus, recommending cheaper parts to use exposes them to greater risk, while making them less money. A lot of people get mad at factories for not being more aggressive on keeping the cost down, but if you think about it, you really have to get engaged. You need to get an engineer working with these guys to cost things down because ultimately, it's your bottom line. It's your net profit. You don't just go to China and expect them to do it right.

An ODM can possibly solve that problem by hiring staff dedicated to reducing costs, but then the ODM would either need to charge the customer extra to make the service sustainable, or require a significantly larger order volume over which to amortize the extra cost of providing such services.

More interoperability in the industry would be good, too. One startup I work with is Spark,* which really tries to enable people to use its hardware platform by being open. I feel like one piece missing for Spark is getting ODMs to be "Spark certified" to make products that use Spark's platform. Often, someone wants to design one product into another product, and suggestions about how to do that effectively are all over

* Eventually, Spark changed its name to *Particle*.

the place. Even if you have all the necessary information, it's not a streamlined process for most people.

When someone is given all the design answers, a lot of decoding still has to happen. Even bigger companies are afraid of that because they don't have the competency to hire the people to get that decoding done.

What is your current focus in the hardware industry?

Right now, I'm working with Jie Qi on circuit stickers. We're getting to the point of shipping the units out, and I'm hell-bent on making sure that I meet the deadlines I set for my campaign. I actually want to ship on time and get things to people when I said I would because there has been way too much lateness in crowdfunded campaigns. It doesn't have to be that way. You just have to set expectations, have your stuff together before you announce the date, and know when the inventory is pretty much ready to go. We have a number of product lines that are selling; about half are done with manufacturing and are just waiting in the factory to ship. A couple of new lines are behind, but we still have until May to solve these issues. I think it will be no problem, and I'm looking forward to seeing our lines grow and develop and work with more people.

The other thing I am working on is this Novena laptop project with Sean Cross, which we weren't really planning on doing last year. I built this handmade prototype last December; it was a little, kind of crummy, leather-and-paper thing. We used it to give a presentation at CCC, and the response was overwhelming. That was great, and I refactored the design to make it more manufacturable and more sourceable. The campaign seems to be going well so far. I think it will fund, and I'm looking forward to getting Novena manufactured and out in the world.

What have you learned from your two crowdfunding campaigns?

Completing almost two crowdfunding campaigns has given me a lot of insight. Earlier, I mentioned that people selling online price their product too low to later move into retail. But it's been really painful to maintain the high price that I say that everyone else should maintain. It's so tempting to go lower to an unsustainable point.

The reason a lot of crowdfunding campaigns fail to deliver is because they price too low. They can't actually build the product for the price they set. Even knowing this, I still had to grit my teeth on the laptop because I had to price it higher than I would have liked. Despite the high price, if we were to close the campaign at exactly the amount I hope to raise, I would probably just barely not lose money on it, but a lot of people don't see that. Look at something like the Ubuntu Edge, which raised $12 million but needed $25 million to succeed. That's because in order to set a price of $700–800 per phone, they had to build 40,000 phones. So even though people thought the Ubuntu Edge was cool and it raised a lot of money, it didn't reach its funding goal, which is a sad conclusion for everyone.

I knew I could either price my laptop much lower and need thousands of people to buy it to reach my goal, or I could service a really focused market of a few hundred open source enthusiasts who are totally on the same page as me. At the end of the day, especially in the early phases, you really want those enthusiasts. They're going to be your best users. You want to take care of them and give them the best service possible. You're going to charge a little more, but you're going to build a really good product for them and they're going to be happy. That's a much happier conclusion in my mind than trying to shoot the moon and failing.

epilogue

When I start hacking or making, it's driven by curiosity. Only a small portion of my work ends up being relevant or interesting, but I journal my successes and my failures at my blog, *http://bunniestudios.com/*, and I occasionally tweet observations at @bunniestudios. It's hard to know what will be a hit or a miss; but as long as I'm learning, the journey is worthwhile. And so I will keep wandering the electronic frontier . . .

index

A

accessories and packaging, 200–201
adaptations, influenza, 333–335
Akiba, 64–65
all-in-one desktop Novena, 218, 242–243
Amendment 1092 to National Defense Authorization Act, 149–150
American vs. Chinese manufacturing, 35–36
amino acids, 328–329
anisotropic tape, 257–259
antibiotic-resistant superbugs, 342–343
anticounterfeit measures for US military, 149, 154–156
Apple
  Apple II, 207, 326–327, 373
  Foxconn, 18, 20
  quality control, 37
  refinement costs, 202
AppoTech chips, 293
approved vendor list (AVL), 76
Arduino, 213, 360
  Arduino Uno, 104–105, 127
  manufacturing, 44–57
    copper sheets, 46–48
    etching PCBs, 51–53
    PCB pattern, applying to copper, 49–50
    soldermask and silkscreen, 53–54
    testing and finishing, 54–57
artisan engineering, 213
Asanović, Krste, 310–311
Ashby chart, 230
audit logs for test programs, 96
authentic parts, keeping reserve of, 156
automation
  for electronics assembly, 29–31
  test program, 96
  in zipper factory, 67–70
AVL (approved vendor list), 76

B

bacteria
  CRISPRs in, 347–350
  metabolic pathways, 325–327
barcode, embedding in chips, 154
battery board, Novena, 223–224
battery pack, Novena, 243–244
beachhead, building, 315–317
bicycle safety light, 74–75, 79–82
bill of materials (BOM), 74–84
  approved vendor list, 76
  for bicycle safety light, 74–75, 79–82
  change, planning for, 82–84
  extended part numbers, 78–79
  form factor, 77–78
  quotations, 107–108
  tolerance, composition, and voltage specification, 76–77
biology and bioinformatics, 277–278
  comparing H1N1 to computer virus, 327–335
    adaptable influenza, 333–335
    DNA and RNA as bits, 328–330
    hacking swine flu, 331–332
    silver lining, 335
    unique access ports, 330–331
  patching genome, 346–354
    CRISPRs in bacteria, 347–350
    gene drive, 352–354
    human engineering, 351–352
    where to cut genes, 350–351
  personalized genomics, 344–346
  reverse engineering superbugs, 335–344
    antibiotic resistance, 342–344
    O104:H4 DNA sequence, 336–338
    reversing tools for biology, 338–340
    UNIX Shell Scripts, 340–342
BLASTX decompiler, 339–340

Blueprint interview, 372–382
BOM. *See* bill of materials
bonding USB chips to PCBs, 61
booting OS, 321
bootstrapping, 197, 203
boot structure, reverse engineering, 311–315
bottom line, and DFM, 88–91
breakout board for beginners, 241–242
building technology without using it, 23–24
business model, 363

C

capacitors, 12, 76–77
case construction
- chumby, 26–28
- Novena, 233–236

cash flow, Chumby, 193
cell phones
- hacking, 306–324
  - attaching debugger, 317–320
  - beachhead, building, 315–317
  - booting OS, 321
  - building new toolchains, 321–323
  - results, 323–324
  - reverse engineering boot structure, 311–315
  - system architecture, 306–311
- screen replacement, 120–121
- $12 phone, 126–140
  - engineer rights, 135–140
  - from gongkai to open source, 134–135
  - hardware, 128–131

CFT (Cyber Fast Track) initiative, 289
change, planning for and coping with, 82–84
check plots, 268
Chibitronics, 251–274
- background, 251–259
- check plots, 268
- Chinese New Year, impact on supply chain, 272–273
- complications regarding simple requests, 267–268
- delivery, 264–266
- developing new process, 259
- incorrect placement of components, 268–269
- last-minute changes, 271–272
- process capability test, 261–264
- shipping, 273–274
- single points of failure, eliminating, 271
- stencil of sticker patterns, 271–272
- test program, 92–94
- translation issues, 270–271
- visiting factory, 260–261

China. *See also* factories; Shenzhen, China
- Chinese New Year, impact on supply chain, 272–273
- Chinese translation problems, 270–271
- technology growth, 364–366

China Software Developer Network (CSDN) interview, 357–372
- about hardware hackers, 367–372
- about open hardware and maker movement, 358–367

chip-on-board (CoB) technology, 29
chips
- bonding to PCBs, 61
- counterfeit, 143–148. *See also* US military hardware, counterfeit chips in
- decapping, 282–283
- hand-placing on PCBs, 59–61
- SEG Electronics Market, 11–14
- for USB memory sticks, 57–59

chip shooters, 30
Chipworks, 246
chroma keying, 303–304
Chumby, 1–2, 181
- automation in assembly, 29–31
- case production, 26–28
- cash flow, 193
- chumby classic, 183–184

Chumby *(continued)*
- chumby One
  - development of, 184–189
  - trim and finish, 101–104
- connector placement, 25–26
- contracts, 193–205
- counterfeit microSD cards
  - authenticity, 159–160
  - electronic card ID data, 158–159
  - forensic investigation, 160–162
  - gathering data, 162–165
  - summarizing findings, 166–168
  - visible differences, 157–158
- factory testing, 41
- factory tours, 16–17
- hacker-friendly platform, 182–184
- injection molding, 31–34
- interview with Phil Torrone, 189–205
- lessons learned from, 374–375
- margin, 192–193
- merchant buyers, 192
- microphone factory installation, 20–23
- motherboard, 188–189
- NeTV. *See* NeTV
- quality control, 36–39
- remote testing, 39–40
- reverse logistics and returns, 193
- test points, 187–188

circuit stickers, 251–274. *See also* Chibitronics
- background, 251–259
- check plots, 268
- Chinese New Year, impact on supply chain, 272–273
- complications for simple requests, 267–268
- delivery, 264–266
- developing new process, 259
- incorrect placement of components, 268–269
- last-minute changes, 271–272
- process capability test, 261–264
- shipping, 273–274
- single points of failure, eliminating, 271
- stencils of, 271–272
- translation issues, 270–271
- visiting factory, 260–261

Circuit Sticker Sketchbook, 256–257, 267–268

clamshell testing, 54

cloning, 116

CoB (chip-on-board) technology, 29

Coders' Rights Project, 137

COGS (cost of goods sold), 90–92

colors, communicating with operators through, 96

community-enforced IP rules, 124–125

community support for Novena, 247–249

company structure, 202–203

composition, BOM, 76–77

computer virus, comparing H1N1 virus to, 327–335
- adaptability, 333–335
- antibodies, 335
- DNA and RNA as bits, 328–330
- hacking H1N1 virus 331–332
- unique access ports in organisms, 330–331

configuration fuses, 281

contracts, negotiating, 193–205

copper sheets, for PCBs, 46–50

copying, 116

copyrights, 137, 138, 175–177

cosmetic blemishes, 87–88

cost of goods sold (COGS), 90–92

counterfeit goods. *See* fake goods

couriers, 112

coverlay, 260–261

craftspeople, need for, 26–28

CRISPR/Cas system, 347–352

Cross, Sean "xobs", 134–135, 215–216, 289–290. *See also* Novena; SD cards, hacking

crowdfunding, 197–198, 265, 266, 382

Crowd Supply, 250, 264, 265

CrypTech, 248–249

custom battery pack problems, 243–244
Cyber Fast Track (CFT) initiative, 289

D

data display channel (DDC), 304
Debian, 246
debugger, attaching, 317–320
decapping IC, 282–283
decompiler, 339–340
dedicated hardware real-time clock (RTC) module, 238–239
dedication to quality, 20–23
defective units, paying for, 3
delivery of circuit stickers, 264–266
design files, sharing, 363
design for manufacturing (DFM), 84–100. *See also* test program
  bottom line, 88–91
  overview, 85–86
  testing vs. validation, 97–100
  tolerances, 86–88
design process, 105–106
design vocabulary, 101
desktop Novena, 218, 242–243
DFM. *See* design for manufacturing
Digital Millennium Copyright Act (DMCA), 137
direct repeat sequence, 348
direct-to-consumer (DTC) personal genomics, 344–345
disease predictions based on mutations, 345
distribution channel, 196
DIY speakers, 237–238
DMCA (Digital Millennium Copyright Act), 137
DNA, 328–330. *See also* genome
double-shot molds, 103–104
DRAM chips, 12–13
drilling process, PCB boards, 46–48
drug resistance, 338–341
DTC (direct-to-consumer) personal genomics, 344–345

E

ECO (engineering change orders), 82–84
E. coli, 342
EDID (extended display identification data), 304
EDK (embedded development kit), 135
EDM (electrical discharge machine), 33
EFF (Electronic Frontier Foundation), 137
effects stickers, 263
EHEC O104:H4, 335–344
  answering questions with UNIX shell scripts, 340–342
  antibiotic resistance, 342–344
  DNA sequence, 336–338
  reversing tools for biology, 338–340
electrical discharge machine (EDM), 33
electronic card ID data, 158–159
Electronic Frontier Foundation (EFF), 137
electronic tolerances, 86–87
embedded development kit (EDK), 135
enclosure, Novena, 224–227
end-of-life (EOL), 82
engineering change orders (ECO), 82–84
engineering humans, 351–352
engineering samples, 170–172
engineer rights, 135–140
  copyrights, 138
  patents and other laws, 136–137
  programming languages, 138–140
EOL (end-of-life), 82
erasing
  flash memory, 284–285
  memory cards, 298
  security bits, 285–287
etching PCBs, 51–53
e-waste, handling, 155–156
extended display identification data (EDID), 304
extended part numbers, 78–79
external mimicry, 150–151

F

factories, 2–3, 43–44. *See also* quality; *specific factories by name*
automation, 29–31
building technology without using it, 23–24
dedication to quality, 20–23
defective units, paying for, 3
feeding workers, 18–20
injection molding, 31–34
mistakes in manufacturing, 34, 41–42
need for craftspeople, 26–28
partnerships with, 107–113
import duties, 113
ordering more units than proven demand, 112
quotations, 108–111
scrap and yield, 111–112
shipping costs, 112
tips for forming, 107–108
scale in Shenzhen, 17–18
scrap, 152
searching for, 378
skilled workers, 24–26
testing, 41
failure analysis services, 281
failures, learning from, 368–369
Fairchild 74LCX244, 146–147
fake goods, 143–174
chips, well-executed, 143–148
chips in US military hardware, 149–156
anticounterfeit measures, 154–156
types of counterfeit parts, 150–153
US military designs, 153–154
FPGAs, 168–174
incorrect ID codes, 170–172
solutions, 173–174
white screen issue, 168–170
microSD cards, 156–168
authenticity, 159–160
electronic card ID data, 158–159
forensic investigation, 160–162
gathering data, 162–165
summarizing findings, 166–168
visible differences, 157–158
feeding factory workers, 18–20
Feist Publications, Inc. v. Rural Telephone Service Co., Inc., 138
Fernly shell, 315–316, 317–319
Fernvale, 306
attaching debugger, 317–320
beachhead, building, 315–317
booting OS, 321
Frond, 307–308
legal tasks, 134–136
peripheral connectors, 308–309
results, 323–324
reverse engineering boot structure, 311–315
system architecture, 306–311
system diagram, 309
toolchains, building new, 321–323
field programmable gate array. *See* FPGAs
film imaging, 49–50
firmware
in memory cards, 292
Novena, 246–247
five-digit multimeter, 98
flash chips, for USB memory sticks, 57–59
flash memory, erasing, 284–285
flat patterns, 26–28
flex circuits, 252–253
flex PCB factory, 260–261
flow marks, 236
flying head testing, 54
form factor, 77–78
forward bias voltage, 88, 89
founders, suggestions for, 199
Foxconn, 18, 20
FPC (internal flexible printed circuit) header, 238–239

FPGAs (field programmable gate array)
counterfeit, 168–174
incorrect ID codes, 170–172
solutions, 173–174
white screen issue, 168–170
future trends, 212–213
Novena, 239
Freescale/NXP iMX6 CPU, 220
front bezel, Novena, 237–238
fully decapped chips, 282
functionally decapped chips, 282–283
fuzzing, 293

G

gene drive, 352–354
General-Purpose Breakout Board (GPBB), 241–242
genome
disease predictions based on mutations, 345
genotyping, 344–345
patching, 346–354
CRISPRs in bacteria, 347–350
engineering humans, 351–352
gene drive, 352–354
where to cut genes, 350–351
reference, 345–346
genotyping, 344–345
ghost shift, 115, 152
golden samples, 36, 82
gongkai (公开), 117–118, 119–120. *See also* shanzhai
cell phone screen replacement, 120–121
defined, 131–134
vs. kai fang yuan dai ma (开放源代码), 372
$12 phone, 126–140
engineer rights, 135–140
from gongkai to open source, 134–135
hardware, 128–131
GPBB (General-Purpose Breakout Board), 241–242
gray markets, 154

H

H1N1 virus, comparing to computer virus, 327–335
adaptability, 333–335
antibodies, 335
DNA and RNA as bits, 328–330
hacking H1N1 virus 331–332
unique access ports in organisms, 330–331
H5 port, 330
hacker-friendly platform, 182–184
hacker spirit, 371
hacking hardware. *See* hardware hacking
hand-placing chips on PCBs, 59–61
hard drive, choosing, 244–246
hardware hacking, 279–281
CSDN interview about, 367–372
general discussion, 275–278
HDCP-secured links to allow custom overlays, 298–306
of PI C18F1320, 281–289
closer look, 283–284
decapping IC, 282–283
erasing flash memory, 284–285
erasing security bits, 285–287
protecting other data, 287–289
of SD cards, 289–298
potential security issues, 298
resource for hobbyists, 298
reverse engineering microcontroller, 293–297
shanzhai phones, 306–324
attaching debugger, 317–320
beachhead, building, 315–317
booting OS, 321
building new toolchains, 321–323
Fernvale results, 323–324
reverse engineering boot structure, 311–315
system architecture, 306–311
structure of cards, 290–293
hardware startups, 378–380
hash function, 315
HDCP-secured links, hacking, 298–306

health, caring for, 205
heirloom laptops, 210–211
Heirloom Novena, 218, 227–232
  hard drive, 245–246
  mechanical engineering details, 229–232
  wood for enclosure, 228–229
honest finishes, 101
horizontal gene transfer, 343
human engineering, 351–352

I

ID codes, FPGA, 170–172
import duties and licenses, 113
i.MX233, 184
incoming quality control (IQC) guidelines, 160
incorrect placement of components on circuit stickers, 268–269
industrial design, 100–106
  Arduino Uno silkscreen art, 104–105
  chumby One trim and finish, 101–104
  personal design process, 105–106
injection molding
  general discussion, 31–34
  in Novena manufacturing, 233–236
innovation, 359
input networks, 87
intellectual property (IP). *See also* gongkai; shanzhai
  general discussion, 115–118
  Western vs. Chinese models, 131–132
internal flexible printed circuit (FPC) header, 238–239
interoperability, 380
interviews, 357–382
  Blueprint, 372–382
  China Software Developer Network (CSDN), 357–372
    about hardware hackers, 367–372
    about open hardware and maker movement, 358–367
  Make:, 189–205
inventory turning, 196–197
investigating fake microSD cards, 158–159, 160–162
involvement in manufacturing process, 36–39
IP. *See* intellectual property
IQC (incoming quality control) guidelines, 160
Ito, Joi, 264

J

Japan, economic development of, 365
JTAG, 170

K

kai fang yuan dai ma (开放源代码), 372
keystreams, 304–306
Kare, Susan, 39
Kickstarter, 197–198, 377
Kingston microSD cards, 156–168
  authenticity, 159–160
  electronic card ID data, 158–159
  forensic investigation of, 160–162
  gathering data, 162–165
  summarizing findings, 166–168
  visible differences, 157–158
knit lines, 235
Kovan, 169

L

labor costs, 110
laptop Novena, 218
laser imaging, 49
last-minute changes, 271–272
LCA (Linux Conference Australia), 57
LCD bezel, Novena, 226
LEDs, in bicycle safety light, 74–75, 79–82
Li, Xiao, 23–24
LinkIT ONE, MediaTek, 323–324
Linux Conference Australia (LCA), 57
logs for test programs, 96

M

Make: interview, 189–205
MakerBot, 203
maker movement, 358–367
managed NAND system, 186–187
man-in-the-middle (MITM) attacks, 290, 298, 301
manufacturer ID, 158–159
manufacturing. *See* factories
margins
  chumby, 192–193
  factory, 110–111
Master Chao, 26–28
MCM (multichip module), 310
mechanical engineering, Novena, 229–232
mechanical tolerances, 87–88
MediaTek LinkIT ONE, 323–324
MediaTek MT6250DA, 130–131
MediaTek MT6260, 140, 310–311
merchant buyers, 192
metal spiral binding, *Circuit Sticker Sketchbook*, 267–268
microcontroller
  in memory cards, 292
  reverse engineering, 293–297
  test program, 92–94
microphone, chumby, 20–23
microSD cards
  chumby One, 186
  counterfeit, 156–168
    authenticity, 159–160
    electronic card ID data, 158–159
    forensic investigation, 160–162
    gathering data, 162–165
    summarizing findings, 166–168
    visible differences, 157–158
military hardware, counterfeit chips in, 149–156
  anticounterfeit measures, 154–156
  types of counterfeit parts, 150–153
  US military designs, 153–154
minimum order quantity (MOQ), 81
min-max spread, 86–87
mirror-finished plastic, 70–71
mistakes in manufacturing, 34, 41–42
MITM (man-in-the-middle) attacks, 290, 298, 301
MIT Media Lab, 264
monastic design, 100
Moore's law, 206–212, 359
MOQ (minimum order quantity), 81
motherboard
  chumby One, 188–189
  Novena, 221–222, 238–239
Mottweiler, Kurt, 228, 238
multichip module (MCM), 310
mutations, disease predictions based on, 345
Mycoplasma pneumoniae, 325–327
MyriadRF, 248

N

NAND flash chips, 13
National Defense Authorization Act, 149–150
NeTV, 280
  background on HDCP, 300–301
  conceptual diagram of, 303
  development of, 299–300
  FPGA diagram, 305
  goals for, 301
  how it worked, 302–303
  keystream, creating, 304–305
  user overlay content, creating, 303–304
New Balance factory, 17–18
Ng, P.C., 344–345
nonrecurring engineering (NRE) costs, 111
Novena, 133, 215–250
  all-in-one desktop, 218, 242–243
  breakout board for beginners, 241–242
  case construction, 233–236
  community support, 247–249
  custom battery pack, 243–244

Novena *(continued)*
design, 219–227
battery board, 223–224
enclosure, 224–227
motherboard, 221–222
dimensions, 219
DIY speakers, 237–238
firmware, 246–247
front bezel changes, 237–238
hard drive, choosing, 244–246
Heirloom, 218, 227–232
hard drive, 245–246
mechanical engineering details, 229–232
wood for enclosure, 228–229
injection molding, 233–236
laptop, 218
motherboard, 238–239
power pass-through board, 242–243
pricing, 218
PVT2 mainboard, 238–240
users, 217–218
NRE (nonrecurring engineering) costs, 111
NuttX, 141

O

O104:H4 DNA sequence, 336–338
ocean freight, 273–274
ODMs (original design manufacturers), 379–380
online hardware startups, 378–380
on-time delivery, 266
open BOM, 124–125
open source, 117, 134–135
hardware, 176–178, 205–214. *See also* Chibitronics; Chumby; Fernvale; Kovan; NeTV; Novena
CSDN interview about, 358–367
heirloom laptops, 210–211
monetization, 195–196
opportunities for, 211–214
trends in, 206–209
software, 362
ordering more units than proven demand, 112
original design manufacturers (ODMs), 379–380
overlay, creating, 303–304
overmolding, 34

P

package type, 77–78
pad printing, 102
palindromic sequences, 348
PAM (proto-space adjacent motif), 350–351
Particle's Spark Core, 306–307
partnerships with factories, 107–113
import duties, 113
order more units than proven demand, 112
quotations, 108–111
scrap and yield, 111–112
shipping costs, 112
tips for forming, 107–108
part numbers, 78–79
patching genome, 346–354
CRISPRs in bacteria, 347–350
engineering humans, 351–352
gene drive, 352–354
where to cut genes, 350–351
patents, 136–137, 194–195
patterning, 46
pattern makers, 26–28
PB2 influenza gene, 331–332
PCBs, 44–57
applying pattern to copper, 49–50
bonding chips to, 61
for circuit stickers, 260–261
copper sheets, 46–48
etching, 51–53
Fernvale Frond, 307–308
hand-placing chips on, 59–61
soldermask and silkscreen, 53–54
testing and finishing, 54–57
PCH China Solutions, 17, 37
Peek, Nadya, 226
Peek array, 226

penicillin resistance, 338–339
Perrott, Joe, 27
personal design process, 105–106
personalized genomics, 344–346
Phase Locked Loop (PLL), 140
photoresist, 49–50
physical identifiers, embedding, 154–155
physical programming, 263
PIC18F1320, hacking, 281–289
  closer look at, 283–284
  decapping IC, 282–283
  erasing flash memory, 284–285
  erasing security bits, 285–287
  protecting other data, 287–289
plastic finishes, 70–71
PLL (Phase Locked Loop), 140
poison pills, 136–137
polyimide, 260–261
power pass-through board, 242–243
pragmatic design, 100
precision, 31–34
pricing
  aiming high, 199–200
  Novena, 218
  quality control, 34–35
probe card, 58
process capability test, 261–264
process geometry, 144–145
production candidate stickers, 263
programming languages, 138–140
protecting data when hacking, 287–289
protein database, 338–339
proteins, 329, 337
proto-space adjacent motif (PAM), 350–351

Q

QC (quality control) room, 36–39
QEMU, 317–318
Qi, Jie, 253–256, 263–264, 270–271. *See also* Chibitronics
quality, 34–35
  American vs. Chinese manufacturing, 35–36
  dedication to, 20–23
  factory testing, 41
  involvement in manufacturing process, 36–39
  mistakes, 41–42
  remote testing, 39–40
quality control (QC) room, 36–39
quaternary structure, 350
quotations, evaluating, 108–111

R

Radio Electronics (无线电), 369
Raspberry Pi, 360
read-evaluate-print-loop (REPL) shell, 293–297
real-time clock (RTC) module, 238–239
reballing, 155
rebinned parts, 151–152
recycling, 154–155
red ring of death, 42
reference genome, 345–346
refurbished parts, 150–151, 154
remote testing, 39–40
repair culture, 213
REPL (read-evaluate-print-loop) shell, 293–297
resistive current limiting, 88
resistors, 76
Restriction of Hazardous Substances (RoHS) testing, 41
retailers, engaging, 200, 378
returns, in retail, 193
reverse engineering, 137
  boot structure, 311–315
  general discussion, 275–278
  microcontroller, 293–297
  superbugs, 335–344
    antibiotic resistance, 342–344
    O104:H4 DNA sequence, 336–338
    reversing tools, 338–340
    UNIX shell scripts, 340–342
reverse logistics, 193
RNA, 328–330
RNA-dependent RNA polymerase, 333

robotics controller, 78
RoHS (Restriction of Hazardous Substances) testing, 41
ROM, dumping, 312–316
rooting, user, 370
routing PCBs, 55
RTC (real-time clock) module, 238–239
rubberized tags, 25

S

Samsung microSD cards, 163–168
SanDisk microSD cards, 163–168
satin-finished plastic, 70–71
scale in factories, 17–18
scarcity and demand, 70–71
Scarmagno, Italy, 44–45
scrap, handling, 111–112
scriptic language, 139–140
SD cards, hacking, 289–298
  potential security issues, 298
  resource for hobbyists, 298
  reverse engineering microcontroller, 293–297
  structure of cards, 290–293
  vulnerabilities, 290
secondary structure, 349–350
second-sourcing, 153
security bits, erasing, 285–287
security issues, SD cards, 298
semiautomated process, in zipper factory, 68–70
sensor and microcontroller stickers, 263
shanzhai (山寨), 116–117, 121–125, 177, 371–372. *See also* gongkai
  cell phones, 2
  community-enforced IP rules, 124–125
  hacking phones, 306–324
    attaching debugger, 317–320
    beachhead, building, 315–317
    booting OS, 321
    building new toolchains, 321–323
    Fernvale results, 323–324
    reverse engineering boot structure, 311–315
    system architecture, 306–311
  more than copycats, 123–124
sharing design files, 363
Shenzhen, China, 1–4. *See also* factories
  screen replacement, 120–121
  SEG Electronics Market, 8–14
  shanzhai organizations in, 123
Shenzhen Bookstore, 14–15
"ship or die" motto, 198–199
shipping products, 112, 273–274
side-by-side bonding, 166
signatures, in memory, 319–320
silkscreen, 53–54, 57
single nucleotide polymorphisms (SNPs), 345–346
single points of failure, eliminating, 271
sink marks, 235
skilled workers, 24–26
smartcards, 144–145
smart watches, 124
SMT (surface mount technology), 55, 77–78
SNPs (single nucleotide polymorphisms), 345–346
soldermask, 53–54, 57
Song Jiang, 122
smartphones. *See* cell phones
spacers, 348
speakers, Novena, 237–238
SPI ROMulator FPGA, 313
ST19CF68 chips, 144–148
stacked CSPs, 166
standardization of platforms, 212
stencil of circuit sticker patterns, 271–272
superbugs, reverse engineering, 335–344
  antibiotic resistance, 342–344
  O104:H4 DNA sequence, 336–338
  reversing tools, 338–340
  UNIX shell scripts, 340–342

supply chain, impact of Chinese New Year on, 272–273
surface mount technology (SMT), 55, 77–78
swine flu. *See* H1N1 virus, comparing to computer virus
switches
  Novena, 237
  validating, 98–99
system architecture, 306–311
System Elettronica, 44–57
  applying PCB pattern to copper sheet, 49–50
  applying soldermask and silkscreen, 53–54
  copper sheets, 46–48
  etching PCBs, 51–53
  testing and finishing, 54–57
System-on-Chip devices, 310–311

T

tampo printing, 102
technology level, in China, 364–366
Tek MDO4104B-6 oscilloscope, 313
tertiary structure, 350
testing
  flash chips, 58–59
  PCBs, 54–57
  vs. validation, 97–100
test jigs, 99–100, 271
test points, chumby One, 187–188
test program, 91–95
  guidelines for, 94–97
  icons, communicating with operators through, 96
  real-world, 92–94
  setup of, 95–96
  update mechanisms for, 97
3D transistors, 245
through-hole packages, 77–78
tolerances, 76–77, 86–88
Tomlin, Steve, 39, 299
toolchains, building new, 321–323
tooling, 233–234
Torrone, Phil, 189–205
toy factories, 29–30
transistor scaling, 210–211
translation problems, 270–271
transparency in factory relationships, 107–108
trim and finish, chumby, 101–104
triple-reassortant virus, 334–335
$12 phone, 126–140
  engineer rights, 135–140
  from gongkai to open source, 134–135
  hardware, 128–131

U

U-Boot (Universal Bootloader), 246
Ubuntu Edge, 382
unique access ports, in organisms, 330–331
Universal Protein Resource (UniProt), 338–339, 341
UNIX shell scripts, answering biological questions with, 340–342
upstreaming, 246
USB flashing tool, open version of, 320–322
USB memory stick factory, 57–64
  beginning of USB sticks, 57–59
  bonding chips to PCBs, 61
  close look at USB stick boards, 61–64
  hand-placing chips on PCBs, 59–61
USB ports, Novena, 237
US military hardware, counterfeit chips in, 149–156
  anticounterfeit measures, 154–156
  types of counterfeit parts, 150–153
  US military designs, 153–154
UV dye in chips, 154–155
UV-erasable programmable read-only memory (UV-EPROM), 284–285, 286

V

vacuum-tube radio schematic, 207
validation vs. testing, 97–100
Vanchip VC5276, 130
Vasut, Marek, 246, 248
venture capitalist funding, 195–196, 197–199
vibrapots, 67–68
viruses. *See* H1N1 virus, comparing to computer virus
V-NAND, 245, 246
voltage specification, BOM, 76–77

W

Wang, Chris "Akiba", 64–65
waste, handling, 155–156
white screen issue, 168–170
wire bonding, 29–30, 61
wood enclosure for Novena, 228–229

X

Xbox 360, 42
Xbox One, 369
Xilinx, 170–174
xobs, 134–135, 215–216, 289–290. *See also* Novena; SD cards, hacking

Y

yield, 84–85, 90, 111–112
Young's modulus, 229–230

Z

zipper factory, 64–71
    fully automated process, 67–68
    irony of scarcity and demand, 70–71
    semiautomated process, 68–70
Z-tape, 257–259

ABOUT THE AUTHOR

Andrew "bunnie" Huang has always had trouble getting up before noon. That, compounded with his tendency to question authority means he will never hold a job at a Fortune 500. Thus, he is grateful for all the beers that he's received from crowdfunding because it means he can get some calories through hydration.